Worldwide Distributor:

RUSH Publications and Educational Consultancy, LLC
1901 60th Place, L7432, Bradenton
FL 34203
United States

All inquiries, suggestions and comments should be addressed to:

ARMA Publishing and Trade Ltd.
Ferhat S Bozbey A Daire 1 Rumeli Hisarustu
Istanbul, 34340
TURKEY

e-mail : meylani@superonline.com
phone : +90 212 263 5742, +90 212 263 5743
mobile : +90 533 363 5287, +90 506 541 8729
fax   : +90 533 982 6740

ISBN: 0974886874

Author: Ruşen MEYLANİ
Cover design by: Pınar ERKORKMAZ; **e-mail:** mail@pinare.com

All graphics and text based on TI 83 – 84 have been used with the permission granted by Texas Instruments.

ACKNOWLEDGMENTS

I would like to thank the students from USA, Turkey and other countries for their helpful comments; I have made sure that in this edition we have observed them all.

I would like to thank Mustafa Atakan ARIBURNU, my ex-student and my lifetime friend for believing in me and for his continuous support.

I would like to thank TEXAS INSTRUMENTS for providing scientists and mathematicians such powerful hand - held computers, the TI family of graphing calculators. With these wonderful machines, teachers of mathematics can go beyond horizons without the need to reinvent the wheel all the time. I would also like to thank TEXAS INSTRUMENTS for providing me with a limited copyright to use the graphs that have been produced by the TI 83 / 84 Family of graphing calculators throughout this book.

I would like to thank Erdel VIDORI of TEXAS INSTRUMENTS for his suggestions on the organization and title of this book as well as his invaluable efforts in establishing the link between myself and TEXAS INSTRUMENTS.

I would like to thank Zeynel Abidin ERDEM, the chairman of the Turkish – American Businessmen Association (TABA) for his valuable support and contributions on our projects.

I would like to thank Emel UYSAL and Seda EREN for their valuable contributions.

I would like to thank Pınar ERKORKMAZ for her excellent work on the cover design.

I would like to thank Yorgo İSTEFANOPULOS, Ayşın ERTÜZÜN, Aytül ERÇİL, Bayram SEVGEN, Zeki ÖZDEMİR, and Nuran TUNCALI for whatever I know of analytical thinking.

I would like to thank my mother and my father for being who I am.

I would like to thank my mother in law and my father in law for their continuous support.

Last but not the least, I would like to express my most sincere gratefulness to my wife who has continuously encouraged and supported me throughout every stage of preparing this book. Therefore I dedicate this book to her.

To My Beloved Wife Ebru…

"Seni Seviyorum!"

PREFACE

If you give yourself to mathematics, mathematics will give you the world.

The tests in this book are harder than the actual tests as they are produced to prepare you better for the real test; they **are not realistic in terms of the level of difficulty of the questions**. Please note that if your intention is to get prepared for the real test, this book is the one; but if you are in search of a book that will make you feel good only, you should look elsewhere.

However, we still claim that **the tests in this book are realistic in the sense that they cover absolutely every type of question which has ever appeared or is likely to appear in the real test**. The 750 questions in this book are what you need for a perfect score or for a significant increase in your score in a short period of time.

We have modified the 15 tests so that **the tests in this edition are in accordance with the latest trends in the SAT Math Level 2 Subject Test**. You should also notice that each question is unique; even the ones that might look similar are different versions of the same question type, deliberately placed in the book for the best prep.

A smart test taker who wishes to use this book **should go over each and every question and its solution** without worrying about the time she or he spends. If a student can get a raw score of 32 or more in a test, then he/she is very close to the perfect score in the actual test. One can comfortably add 100 to her/his scaled score in order to estimate what she/he will get in the real test.

Moreover, we have made sure in this edition that the errors or ambiguities related to the use of the English language are eliminated and that each question clearly explains what is given and what is to be found. However we apologize for any errors or vague expressions that might still exist. Please do note that this is a math book and not an English grammar book.

Consequently, **this very book is prepared by experts of SAT Mathematics who intend to help college bound SAT takers in improving their scores in the quickest and the best way possible**.

So, best of luck and be prepared…

Ruşen MEYLANİ

TABLE OF CONTENTS

CHAPTER 1

WHAT SAT MATH SUBJECT TESTS

ARE ALL ABOUT

Blank Page

Mathematics Level 1 and Mathematics Level 2 are the two subject tests that the College Board offers. Both tests require at least a scientific, preferably a graphing, calculator. Each test is one hour long. These subject tests were formerly known as the Math Level IC and Math Level IIC subject tests.

Mathematics Level 1 Subject Test

Structure

A Mathematics Level 1 test is made of 50 multiple choice questions from the following topics:

- Algebra and algebraic functions
- Geometry (plane Euclidean, coordinate and three-dimensional)
- Elementary statistics and probability, data interpretation, counting problems, including measures of mean, median and mode (central tendency.)
- Miscellaneous questions of logic, elementary number theory, arithmetic and geometric sequences.

Calculators in the Test

Approximately 60 percent of the questions in the test should be solved without the use of the calculator. For the remaining 40 percent, the calculator will be useful if not necessary.

Mathematics Level 2 Subject Test

Structure

A Mathematics Level 2 test also is made of 50 multiple choice questions. The topics included are as follows:

- Algebra
- Geometry (coordinate geometry and three-dimensional geometry)
- Trigonometry
- Functions
- Statistics, probability, permutations, and combinations
- Miscellaneous questions of logic and proof, elementary number theory, limits and sequences

Calculators in the Test

In Math Level 2, 40 percent of the questions should be solved the without the use of the calculator. In the remaining 60 percent, the calculator will be useful if not necessary.

Which calculator is allowed and which is not:

The simplest reference to this question is this: No device with a QWERTY keyboard is allowed. Besides that any hand held organizers, mini or pocket computers, laptops, pen input devices or writing pads, devices making sounds (Such as "talking" computers) and devices requiring electricity from an outlet will not be allowed. It would be the wisest to stick with TI 84 or TI 89. Both of these calculators are easy to use and are the choices of millions of students around the world who take SAT exams and also university students in their math courses. It is very important to be familiar with the calculator that you're going to use in the test. You will lose valuable time if you try to figure it out during the test time.

Be sure to learn to solve each and every question in this book. They are carefully chosen to give you handiness and speed with your calculator. You will probably gain an extra 150 to 200 points in a very short period of time.

IMPORTANT: Always take the exam with fresh batteries. Bring fresh batteries and a backup calculator to the test center. You may not share calculators. You certainly will not be provided with a backup calculator or batteries. No one can or will assist you in the case of a calculator malfunction. In such case, you have the option of notifying the supervisor to cancel your scores for that test. Therefore, always be prepared for the worst case scenario (Don't forget Murphy's Rules.)

Number of questions per topics covered

The following chart shows the approximate number of questions per topic for both tests.

Topics	Approximate Number of Questions	
	Level 1	Level 2
Algebra	15	9
Plane Euclidean Geometry	10	0
Coordinate Geometry	6	6
Three-dimensional Geometry	3	4
Trigonometry	4	10
Functions	6	12
Statistics	3	3
Miscellaneous	3	6

Similarities and Differences

Some topics are covered in both tests, such as elementary algebra, three-dimensional geometry, coordinate geometry, statistics and basic trigonometry. But the tests differ greatly in the following areas.

Differences between the tests

Although some questions may be appropriate for both tests, the emphasis for Level 2 is on more advanced content. The tests differ significantly in the following areas:

Geometry

Euclidian geometry makes up the significant portion of the geometry questions in the Math Level 1 test. Though in Level 2, questions are of the topics of coordinate geometry, transformations, and three-dimensional geometry and there are no direct questions of Euclidian geometry.

Trigonometry

The trigonometry questions on Level 1 are primarily limited to right triangle trigonometry and the fundamental relationships among the trigonometric ratios. Level 2 places more emphasis on the properties and graphs of the trigonometric functions, the inverse trigonometric functions, trigonometric equations and identities, and the laws of sines and cosines. The trigonometry questions in Level 2 exam are primarily on graphs and properties of the trigonometric functions, trigonometric equations, trigonometric identities, the inverse trigonometric functions, laws of sines and cosines. On the other hand, the trigonometry in Level 1 is limited to basic trigonometric ratios and right triangle trigonometry.

Functions

Functions in Level 1 are mostly algebraic, while there are more advanced functions (exponential and logarithmic) in Level 2.

Statistics

Probability, mean median, mode counting, and data interpretation are included in both exams. In addition, Level 2 requires permutations, combinations, and standard deviation.

In all SAT Math exams, you must choose the best answer which is not necessarily the exact answer. The decision of whether or not to use a calculator on a particular question is your choice. In some questions the use of a calculator is necessary and in some it is redundant or time consuming. Generally, the angle mode in Level 1 is degree. Be sure to set your calculator in degree mode by pressing "Mode" and then selecting "Degree." However, in Level 2 you must decide when to use the "Degree" mode or the "Radian" mode. There are figures in some questions intended to provide useful information for solving the question. They are accurate unless the question states that the figure is not drawn to scale. In other words, figures are correct unless otherwise specified. All figures lie in a plane unless otherwise indicated. The figures must NOT be assumed to be three-dimensional unless they are indicated to be. The domain of any function is assumed to be set of all real numbers x for which f(x) is a real number, unless otherwise specified.

Important Notice on the Scores

In Level 1 questions the topics covered are relatively less than those covered in the Level 2 test. However, the questions in the Level 1 exam are more tricky compared to the ones in Level 2. This is why if students want to score 800 in the Level 1 test, they have to answer all the 50 questions correctly. But in the Level 2 test, 43 correct answers (the rest must be omitted) are sufficient to get the full score of 800.

Scaled Score	Raw Score in Level 1 Test	Raw Score in Level 2 Test
800	50	43
750	45	38
700	38	33
650	33	28
600	29	22
550	24	16
500	19	10
450	13	3
400	7	0
350	1	-3

CHAPTER 2

MOST ESSENTIAL GRAPHING

CALCULATOR TECHNIQUES

COURTESY   TEXAS INSTRUMENTS

Blank Page

COURTESY TEXAS INSTRUMENTS

In this section you will learn the most common graphing calculator techniques along with some critical examples that will raise your scores by at least 50 points. However, there are a lot more to what can be done with the graphing calculator during the test for a typical raise in the score by 150 to 200 points in a very short period of time. Here is a partial snapshot of what you can use your graphing calculator for:

- Polynomial Equations
- Algebraic Equations
- Absolute Value Equations
- Exponential and Logarithmic Equations
- System of Linear Equations, Matrices and Determinants
- Trigonometric Equations
- Inverse Trigonometric Equations
- Polynomial, Algebraic and Absolute Value Inequalities
- Trigonometric Inequalities
- Maxima and Minima
- Domains and Ranges
- Evenness And Oddness

- Graphs of Trigonometric Functions
  - Period
  - Frequency
  - Amplitude
  - Offset
  - Axis of wave equation
- Miscellaneous Graphs
- The Greatest Integer Function
- Parametric Graphing
- Polar Graphing
- Limits and Continuity
- Horizontal and Vertical Asymptotes
- Complex Numbers
- Permutations and Combinations

For learning the topics listed above please refer to the following book which is one of a kind:

SAT Math Subject Test with the TI 83 – 84 Family

What is special about the method in this book is that it shortens

40 hours of college preparatory precalculus study **to** an easy **4 hours.**

P.S. We highly recommend the usage of a TI 84 or a TI 89 graphing calculator.

COURTESY ✦ TEXAS INSTRUMENTS

Solving Equations

When solving a polynomial or algebraic equation in the form **f(x)=g(x),** perform the following steps:

i. Write the equation in the form: **f(x)-g(x)=0**.

ii. Plot the graph of **y=f(x)-g(x)**.

iii. Find the x-intercepts using the **Calc Zero** of TI-84 Plus. However when the graph seems to be tangent to the x-axis at a certain point, you may use the **Calc Min** or **Calc Max** facilities but you should make sure that the y-coordinate of the minimum or maximum point is zero.

Solving Inequalities

When solving an inequality in the form **f(x)<g(x),** or **f(x)≤g(x),** or **f(x)>g(x),** or **f(x)≥g(x)** perform the following steps:

i. Write the inequality in the form: **f(x)-g(x)<0** or **f(x)-g(x)≤0** or **f(x)-g(x) >0** or **f(x)-g(x)≥0.**

ii. Plot the graph of **y=f(x)-g(x)**.

iii. Find the x-intercepts using the **Calc Zero** of TI-84 Plus. However when the graph seems to be tangent to the x-axis at a certain point, you may use the **Calc Min** or **Calc Max** facilities but you should make sure that the y-coordinate of the minimum or maximum point is zero.

iv. Any value like **-6.61E -10** or **7.2E -11** can be interpreted as 0 as they mean **-6.6x10** $^{-10}$ and **7.2x10** $^{-11}$ respectively.

v. The solution of the inequality will be the set of values of x for which the graph of f(x)-g(x) lies below the x axis if the inequality is in one of the forms **f(x)-g(x)<0** or **f(x)-g(x)≤0.** The solution of the inequality will be the set of values of x for which the graph of f(x)-g(x) lies above the x axis if the inequality is in one of the forms **f(x)-g(x)>0** or **f(x)-g(x)≥0.** If ≤ or ≥ symbols are involved, then the x-intercepts are also in the solution set.

vi. Please note that the x-values that correspond to asymptotes are never included in the solution set.

Example 1:

$P(x)= 2x^2+3x+1$

$P(a)= 7 \Rightarrow a=?$

Solution:

$2a^2+3a+1=7$

$2a^2+3a-6=0$

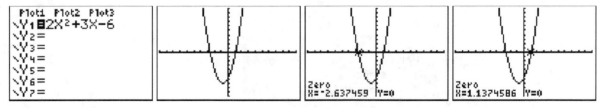

Answer: -2.637 or 1.137

Example 2:

$f(x)= \sqrt{3x+4}$ and $g(x)= x^3$; If is given what (fog)(x)=(gof)(x), then what is x?

COURTESY   **TEXAS INSTRUMENTS**

Solution:

$(f \circ g)(x) = \sqrt{3x^3 + 4}$ and $(g \circ f)(x) = (\sqrt{3x + 4})^3$

$\sqrt{3x^3 + 4} = (\sqrt{3x + 4})^3 \Rightarrow \sqrt{3x^3 + 4} - (\sqrt{3x + 4})^3 = 0$

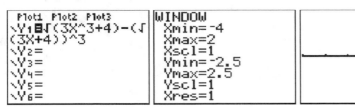

Answer: -1

Example 3:

$|x-3| + |2x+1| = 6 \Rightarrow x = ?$

Solution:

$|x-3| + |2x+1| - 6 = 0$

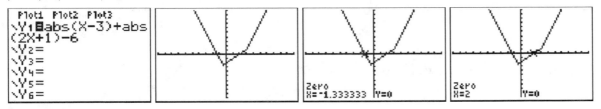

Answer: $x = -1.33$ or 2

Example 4:

$2^{x+3} = 3^x \Rightarrow x = ?$

Solution:

$2^{x+3} - 3^x = 0$

Answer: 5.129

Example 5:

$3.281^x = 4.789^y \Rightarrow \dfrac{x}{y} = ?$

Solution:

If $y = 1$ then $\dfrac{x}{y} = x \Rightarrow 3.281^x = 4.789^1 = 4.789$

COURTESY **TEXAS INSTRUMENTS**

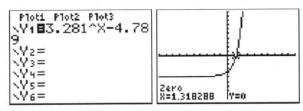

Answer: 1.318

Example 6:

$\log_x 3 = \log_4 x \Rightarrow$ What is the sum of the roots of this equation?

Solution:

$\log_x 3 - \log_4 x = 0$

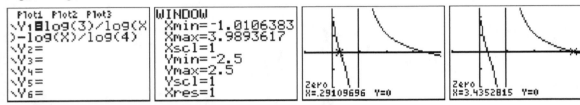

Answer: 0.291+3.435=3.726

Example 7:

$\cos(2x) = 2\sin(90°-x)$. What are all possible values of x between 0° & 360°?

Solution:

Mode: Degrees

$\cos(2x) - 2\sin(90°-x) = 0$

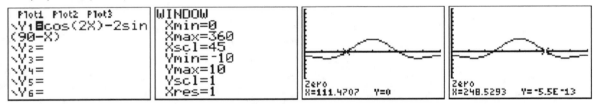

Answer: 111.47°, 248.53°

Example 8:

$2\sin x + \cos(2x) = 2\sin^2 x - 1$ and $0 \le x < 2\pi \Rightarrow x = ?$

Solution:

Mode: Radians

$2\sin x + \cos(2x) - 2\sin^2 x + 1 = 0$

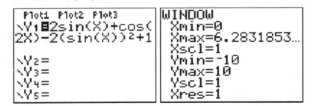

COURTESY ✦ TEXAS INSTRUMENTS

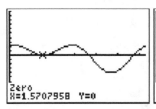 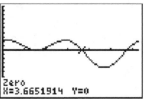 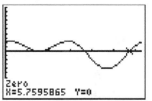

Answer: 1.57, 3.67, 5.76

Example 9:

Solve for x: $\cos^{-1}(2x - 2x^2) = \dfrac{2\pi}{3}$

Solution:

Mode: Radians

$\cos^{-1}(2x - 2x^2) - \dfrac{2\pi}{3} = 0$

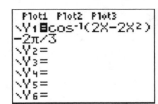 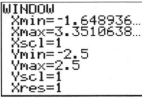 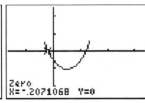 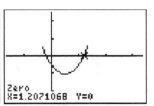

Answer: -0.207 or 1.207

Example 10:

Solve for x: $\dfrac{|x-2|}{x} > 3$

Solution:

$\dfrac{|x-2|}{x} - 3 > 0$

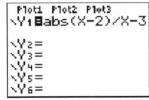 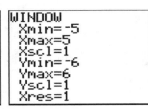 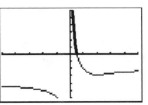

Answer: (0, 0.5)

Example 11:

Solve for x: $|x - 2| \le 1$

Solution:

$|x - 2| - 1 \le 0$

Answer: [1, 3]

COURTESY **TEXAS INSTRUMENTS**

Example 12:

$x^2(x-2)(x+1) \geq 0$

Solution:

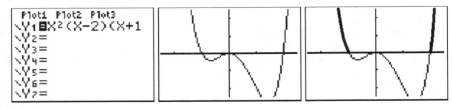

Answer: $(-\infty, -1]$ or $\{0\}$ or $[2, \infty)$

Example 13:

$\sin(2x) > \sin x$

Find the set of values of x that satisfy the above inequality in the interval $0 < x < 2\pi$.

Solution:

$\sin(2x) - \sin x > 0$

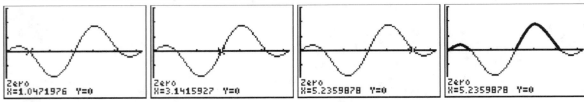

Answer: $(0, 1.05)$ or $(3.14, 5.24)$

Example 14:

$\cos(2x) \geq \cos x$

Find the set of values of x that satisfy the above inequality in the interval $0 \leq x \leq 360°$.

Solution:

$\cos(2x) - \cos x \geq 0$

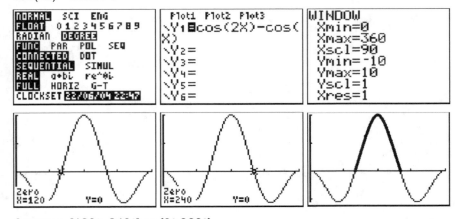

Answer: $[120°, 240°]$ or $\{0°, 360°\}$

COURTESY ✦ **TEXAS INSTRUMENTS**

Miscellaneous Graphing Questions

<u>**Example 15:**</u>

What is the amplitude, period, frequency, axis of wave and offset of y=5sin(x)+12cos(x)-2?

Solution:

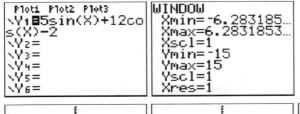

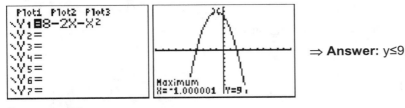

Answer: Amplitude = (11+15) / 2 = 13; Offset: (11 –15) / 2 = -2; Axis of wave: y = -2

Period = 0.394-(-2.747) = 3.141 = π; Frequency = $1/\pi$

<u>**Example 16:**</u>

Find range of $y=8-2x-x^2$

Solution:

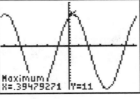

$\Rightarrow$ **Answer:** y≤9

<u>**Example 17:**</u>

Find domain and range of the function given by $y= x^{-4/3}$.

Solution:

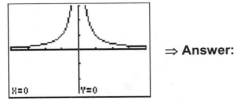

$\Rightarrow$ **Answer:** Domain: x≠0; Range: y>0.

<u>**Example 18:**</u>

Find domain and range of $y= \sqrt{x^2 - 9}$.

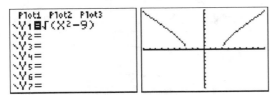

Answer: Domain: x ≤ -3 or x ≥ 3; range: y ≥ 0.

COURTESY **TEXAS INSTRUMENTS**

Example 19:

What happens to sinx as x increases from $-\frac{\pi}{4}$ to $\frac{3\pi}{4}$?

Solution:

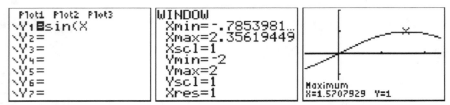

Answer: The function increases between $\pi/4$ and $\pi/2$ and then decreases between $\pi/2$ and $3\pi/4$.

Example 20:

Find the point of intersection of the graphs y=logx and y= $\ln\frac{x}{2}$

Solution:

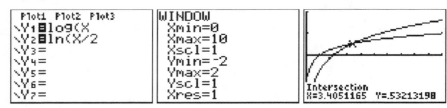

Answer: (3.41, 0.53)

Example 21:

f(x)=2x²+12x+3. If the graph of f(x-k) is symmetric about the y axis, what is k?

Solution:

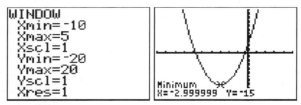

Answer: The graph must be shifted 3 units toward right therefore k=3.

Example 22:

f(x)= -(x-1)²+3 and -2 ≤ x ≤ 2. Find the range of f(x).

Solution:

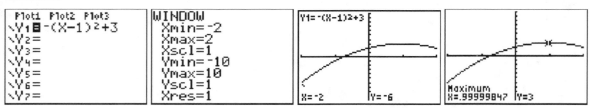

Answer: -6 ≤ y ≤ 3.

COURTESY TEXAS INSTRUMENTS

CHAPTER 3

MODEL TESTS

Blank Page

Model Test 1

Test Duration: 60 Minutes

Directions: For each of the following problems, decide which is the **best** of the choices given. If the exact numerical value is not one of the choices, select the choice that best approximates this value. Then fill in the corresponding oval on the answer sheet.

Notes:

- A calculator will be necessary for answering some (but not all) of the questions in this test. For each question you will have to decide whether or not you should use a calculator. The calculator you use must be at least a scientific calculator; programmable calculators and calculators that can display graphs are permitted.

- For some questions in this test you may have to decide whether your calculator should be in the radian mode or the degree mode.

- Figures that accompany problems in this test are intended to provide information useful in solving the problems. They are drawn as accurately as possible **except** when it is stated in a specific problem that its figure is not drawn to scale.

- All figures lie in a plane unless otherwise indicated.

- Unless otherwise specified, the domain of any function **f** is assumed to be the set of all real numbers **x** for which **f(x)** is a real number.

Reference Information: The following information is for your reference in answering some of the questions in this test.

- Volume of a right circular cone with radius **r** and height **h**: $V = \frac{1}{3}\pi r^2 h$

- Lateral area of a right circular cone with circumference of the base **c** and slant height **l**: $S = \frac{1}{2}cl$

- Volume of a sphere with radius **r**: $V = \frac{4}{3}\pi r^3$

- Surface area of sphere with radius **r**: $S = 4\pi r^2$

- Volume of a pyramid with base area **B** and height **h**: $V = \frac{1}{3}Bh$

1. The number $\log_5 1234$ is between which pair of consecutive integers?

(A) 3 and 4   (B) 4 and 5   (C) 5 and 6   (D) 6 and 7   (E) 7 and 8

2. Given that n is a positive integer, the list of numbers 2n + 3, 2n + 5 and 2n + 7 is called a prime triple if all numbers in the list are prime. How many prime triples are there?

(A) 0   (B) 1   (C) 2   (D) 3   (E) More than 3

GO ON TO THE NEXT PAGE ▶▶▶

Model Test 1

3. Set A contains the prime numbers less than 25. What is the probability that a randomly selected number from set A is less than 8?

(A) $\frac{4}{9}$ (B) $\frac{7}{24}$ (C) $\frac{7}{25}$ (D) $\frac{8}{24}$ (E) $\frac{8}{25}$

$2, 3, 5, 7, 11, 13, 17, 19, 23$

4. For which of the following functions is the inverse not defined?

(A) $f(x) = e^x$ (B) $f(x) = \log x$ (C) $f(x) = -\frac{1}{x}$ (D) $f(x) = x^5$ (E) $f(x) = 7$

5. What is the sum of the zeros of $f(x) = (x - 2)(x^2 - 4)$?

(A) -8 (B) -2 (C) 0 (D) 2 (E) 8

$2, -2, 2$

6. The amount of crops C in a certain year that a field will yield in tons per square meters is a function of the average daily amount of rain R in cubic meters in that year and it is given by $C = \frac{3}{5} + \frac{2R}{7} - \frac{5}{12} \cdot \left(\frac{R}{8}\right)^2$. What is the total amount of crops rounded to the nearest thousand tons that a 20,000 m^2 field will yield when the average daily amount of rain in that particular year was 12 m^3?

(A) 61,000 (B) 62,000 (C) 63,000 (D) 64,000 (E) 65,000

$\frac{f(\pi) - g(0)}{\pi - g(e)} \rightarrow$ 14.7

7. Given that $f(x) = 2x^2 - 2$ and $g(x) = x^3 - 3x^2 + 3$, what is the slope of the line segment AB if points A and B are given by $(\pi, f(\pi))$ and $(g(e), g(0))$ respectively?

(A) 6.62 (B) 6.63 (C) 0.918 (D) 17.73 (E) 17.74

GO ON TO THE NEXT PAGE ▶▶▶

Model Test 1

8. For the function f(x) given f(2) + f(1) = ?

(A) 1   (B) 3   (C) – 1   (D) – 3   (E) – 4

$$f(x) = \begin{cases} \dfrac{3}{x-1} & \text{if } x \neq 1 \\ x^2 - 2x - 3 & \text{if } x = 1 \end{cases}$$

(handwritten) 3 + -4

(handwritten) $5(x^2-1)$ $\dfrac{5(x+1)(x-1)}{(x-1)}$

9. If f(x) = $\dfrac{5x^2 - 5}{x+1}$, then the range of f does not include

(A) 0   (B) 5   (C) -5   (D) -10   (E) 10

(handwritten) 5x = 10 ; x = 2

(handwritten) 5x - 5 = -5 ; 5x = 0 ; x = 0   5x - 5 = -10 ; 5x = -5 ; x = -1   5x - 5 = 0 ; 5x = 5 ; x = 1

10. Among the 120 students in the senior class of **RUSH** Academy, 55 are girls and of the 40 students in the senior class who attend the talent show, 18 are boys. What is the probability that a randomly selected student has attended the talent show given that the student is a girl?

(A) 11/20   (B) 9/20   (C) 11/60   (D) 2/5   (E) 18/65

(handwritten) $\dfrac{55}{120}$ = girls

(handwritten) $\dfrac{18}{40}$ = boys @ talent

(handwritten) $\dfrac{22}{40}$ - girls @ talent

11. How many integer values of x satisfy the inequality given by $-3 < \dfrac{x-2}{x+3} < 4$?

(A) 2   (B) 4   (C) 6   (D) 7   (E) More than 7

(handwritten) $-3(x+3) < x-2 < 4(x+3)$

(handwritten) $-3x - 9 < x - 2 < 4x + 12$

12. What is the measure of the largest angle of triangle ABC given in figure 1?

(A) 33.5°   (B) 33.6°   (C) 62.1°   (D) 84.3°   (E) 93.4°

(handwritten) $-3x < x + 7 < 4x + 21$

(handwritten) $-4x < 7 < 3x + 21$

(handwritten) $0 < 7 + 4x < 7x$

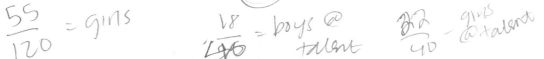

(handwritten) 45   81   72

(handwritten) $c^2 = a^2 + b^2 - 2ab \cos C$

Figure 1

Figure not drawn to scale

GO ON TO THE NEXT PAGE ▶▶▶

Model Test 1

13. Which of the following function – inverse pair is false?

(A) $f(x) = e^{2x}$ and $f^{-1}(x) = \dfrac{\ln x}{2}$

(B) $f(x) = \log(3x)$ and $f^{-1}(x) = \dfrac{10^x}{3}$

(C) $f(x) = \sqrt{4x}$ and $f^{-1}(x) = \dfrac{x^2}{4}$

(D) $f(x) = 1 - x$ and $f^{-1}(x) = 1 - x$

(E) $f(x) = \dfrac{2x+1}{x-3}$ and $f^{-1}(x) = \dfrac{3x+1}{x-2}$

14. An ellipse is given by the relation $f(x, y) = 1$. Which of the following results when the ellipse is shifted 2 units up and 1 unit toward left?

(A) $f(x, 2y) = 1$ (B) $f(x + 1, y - 2) = 1$ (C) $f(x + 1, y + 2) = 1$

(D) $f(x, y/2) = 1$ (E) $f(x - 1, y - 2) = 1$

15. If $3.5 - \sqrt{2}$ and $3.5 + \sqrt{2}$ are the roots of a quadratic equation $ax^2 + bx + c = 0$; then which of the following is not correct?

(A) a is nonzero   (B) discriminant is positive   (C) a, b and c are all real or all complex

(D) $\dfrac{b}{a}$ and $\dfrac{c}{a}$ must be both rational   (E) $\dfrac{b}{a}$ and $\dfrac{c}{a}$ must be both integral

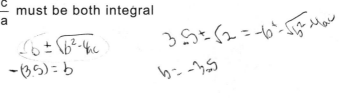

16. If the parabola given by $y = x^2 - kx + 4$ has two real and distinct zeros then what are all possible values of k?

(A) $k < -2$ or $k > 2$ (B) $k < -4$ or $k > 4$ (C) $k \le -4$ or $k \ge 4$ (D) $-2 < k < 2$ (E) $-4 < k < 4$

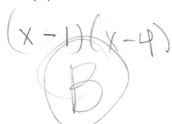

GO ON TO THE NEXT PAGE ▶▶▶

Model Test 1

17. A sphere and a cylinder have the same volumes and radii. What is the ratio of the radius of the sphere to the height of the cylinder?

(A) $\frac{1}{4}$ (B) $\frac{3}{4}$ (C) $\frac{3\pi}{4}$ (D) $\frac{4}{3}$ (E) $\frac{4}{3\pi}$

$$\frac{4}{3}\pi r^3 = \pi r^2 h \qquad \frac{4}{3}r=h \qquad \frac{r}{h}=\frac{3}{4}$$

18. If the right triangle given in figure 2 is rotated about the y axis for 360° what will be the volume of the resulting solid?

(A) 153.93 (B) 153.94 (C) 307.87 (D) 307.88 (E) 461.81

Figure 2

$$y=\frac{3}{7}x$$
$$\pi \int_0^3 7^2-\left(\frac{3}{7}x\right)^2$$

$y=\frac{3}{7}x$ $7y=3x$ $\frac{7}{3}y=x$

19. An object is projected vertically upwards so that t seconds later its height in meters is given by $h(t) = 100 + 20t - 5t^2$. The maximum height of the object is

(A) 40 meters (B) 75 meters (C) 115 meters (D) 120 meters (E) 140 meters

$h'(t)=20-10t$
$0=20-10t$
$+20$ $t=2$

20. In Can's bookshelf there are m mathematics books and p physics books. When Can buys n more mathematics books, half of the books on the shelf will be physics books. Which of the following gives the relation between m, n and p correctly?

(A) m + n = p (B) m − n = p (C) p + n = m (D) p − n = m (E) m + n = 2p

$m+n$ $\frac{p}{p+m+n}=\frac{1}{2}$ $p+m+n=2p$ $m+n=p$

21. If $4 + \sqrt{2}$ is one root of a quadratic equation given by $x^2 − Px + Q = 0$ where P and Q are rational numbers then P is

(A) 0 (B) 4 (C) 8 (D) −4 (E) −8

$(x-4+\sqrt{2})(x-\sqrt{2})=0$

GO ON TO THE NEXT PAGE ▶▶▶

(23)

Model Test 1

22. 9 distinct lines in a plane are given by a || b || c || d and e || f || g || h || m where each small letter represents a different line that does not coincide with the others. What is the maximum number of intersection points of these lines?

(A) 9   (B) 10   (C) 18   (D) 20   (E) 24

23. If sin x + cos x = 1.40, then sinx·cosx =

(A) 0.40   (B) 0.48   (C) 0.96   (D) 1.96   (E) It cannot be determined from the information given.

For questions 24 and 25 please refer to the data given in the following table.

A road with a grade of n% means that the height of a car driving on that road changes by n units for every change of 100 units in horizontal distance. For example a car traveling on a road with 5% grade would mean that the car traveling uphill on this road will rise 5 feet for every horizontal distance of 100 feet it covers.

Road	Grade of the road (%)	Length of the road (mi)
A	6.5	7450
B	5.7	3340
C	9.3	6520
D	4.2	9980
E	8.6	5570

24. What is the angle of elevation of road B?

(A) 2.41°   (B) 3.26°   (C) 3.72°   (D) 4.92°   (E) 5.31°

25. Road D climbs up a maximum height of

(A) 389 mi   (B) 403 mi   (C) 410 mi   (D) 418 mi   (E) 419 mi

GO ON TO THE NEXT PAGE ▶▶▶

Model Test 1

26. For what values of x is the expression $\dfrac{\sin^3 x - \cos^3 x}{\tan^3 x}$ undefined?

(A) $k\pi$ where k is an arbitrary integer

(B) $2k\pi$ where k is an arbitrary integer

(C) $\dfrac{3k\pi}{2}$ where k is an arbitrary integer

(D) $\dfrac{(2k+1)\pi}{2}$ where k is an arbitrary integer

(E) $(2k + 1)\pi$ where k is an arbitrary integer

27. If $y = 5^{-x}$, which of the following gives all possible values of x for which $y > 0$?

(A) $x > 0$ (B) $x < 0$ (C) $x \geq 0$ (D) $x \leq 0$ (E) all real numbers

28. Which of the following can be the equation of the circle given in figure 3?

(A) $x^2 + y^2 - 14x - 6y = 0$

(B) $x^2 + y^2 - 14x - 6y + 58 = 0$

(C) $x^2 + y^2 - 6x - 14y + 58 = 0$

(D) $x^2 + y^2 + 6x - 14y + 42 = 0$

(E) $x^2 + y^2 - 6x - 14y + 42 = 0$

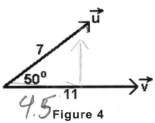

Figure 3

29. What is the magnitude of $\vec{u} + \vec{v}$ for the vectors $\vec{u}$ and $\vec{v}$ given in figure 4?

(A) 8.42   (B) 8.43   (C) 16.4   (D) 71   (E) 269

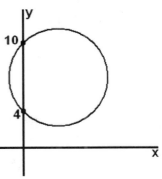

Figure 4

30. Which of the following is the solution set of the system of equations given?

$x^2 + y^2 = 169$
$y = x + 7$

(A) $\{(-13,0),(0,-13)\}$ (B) $\{(-12,-5),(5,12)\}$ (C) $\{(-12,-5),(12,5)\}$

(D) $\{(-5,-12),(5,12)\}$ (E) $\{(0,13),(13,0)\}$

GO ON TO THE NEXT PAGE ▶▶▶

(25)

Model Test 1

31. If $x^9 - 64x^3 = 0$, then which of the following can be x?

(A) – 2   (B) – 4   (C) – 8   (D) 8   (E) 4

32. The function f(x) is given by f(x) = $\dfrac{3x^2 + 5}{7x}$. As x gets infinitely large f(x) approaches

(A) 3x/7   (B) 5/(7x)   (C) 7x   (D) $3x^2 + 5$ (E) 3/(7x)

33. If $x^{\frac{3}{4}} + 2x = 20$, then x can equal

(A) 7.4   (B) 7.5   (C) 7.6   (D) 7.7   (E) 7.8

34. It is given that f(x) is a periodic function whose period is 7. If f(x) has a zero at x = 4 then it has another zero at

(A) 12   (B) 15   (C) 18   (D) 21   (E) 24

35. Which of the following functions has a frequency of $\dfrac{1}{\pi}$?

(A) f(x) = cos(x)   (B) f(x) = cos(2x)   (C) f(x) = 2·cos(x)   (D) f(x) = cos$\dfrac{x}{2}$ (E) f(x) = $\dfrac{\cos(x)}{2}$

GO ON TO THE NEXT PAGE ▶▶▶

Model Test 1

36. The operation given by (a,b)◊(c,d)=(a+bc,bd) is defined on the set of ordered pairs with nonzero second elements. Which of the following is the identity element for this operation?

(A) (0,0)　　　　(B) (1,1)　　　　(C) (1,0)　　　　(D) (0,1)　　　　(E) (0,-1)

37. In figure 5, PB bisects angle QPR. Which of the following equals $\dfrac{a}{b}$?

(A) $\dfrac{q}{r}$　　(B) $\dfrac{\sin R}{\sin Q}$　　(C) $\dfrac{\cos R}{\cos Q}$　　(D) $\dfrac{\tan R}{\tan Q}$

(E) None of the above

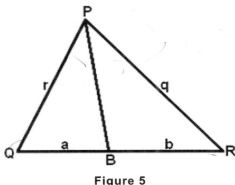

Figure 5

38. It is given that f(x) = ax + b and f(g(x)) = c. If a, b, and c are distinct nonzero real numbers then which of the following is a correct statement for g(x)?

(A) It is a linear function.　　(B) It is a periodic function.　　(C) It is a constant function.

(D) It is a quadratic function.　　(E) It is an exponential function.

$$f(g(x)) = a(g(x)) + b = c$$

39. Which of the following is an asymptote of the function given by f(x) = $3e^{-5x}$ + 7?

(A) x = − 5　　　　(B) x = 3　　　　(C) x = 0　　　　(D) y = 0　　　　(E) y = 7

40. Which of the following is not always correct when 0° < x < 45°?

(A) cos(x) < sec(x)　　(B) sin(x) < csc(x)　　(C) sin(x) < cos(x)　　(D) cos(x) > 0.7　　(E) sin(x) > 0.707

GO ON TO THE NEXT PAGE ▶▶▶

Model Test 1

41. PAB is an isosceles right triangle where P measures 90°. If points A and B are fixed in space, the locus of all such points P is

(A) a line   (B) a point   (C) two points   (D) a circle   (E) a semicircle

42. If f(x) is the least integer greater than or equal to x and $g(x) = -\dfrac{10}{x}$; f(g(3)) + f(g(-6)) = ?

(A) -3   (B) -1   (C) 0   (D) 1   (E) 2

43. A quadratic function is given by h(x) = ax^2 + bx + c where a, b, and c are all nonzero real numbers. The function h(x) intersects the x axis at two distinct points and satisfies the relation h(2−d) = h(2+d) for all real values of d. Which of the following can be determined?

     I. the range of h(x)

     II. sum of the zeros of h(x)

     III. the axis of symmetry of h(x)

(A) I only   (B) II only   (C) III only   (D) II and III only   (E) I, II and III

44. For the circle given in figure 6, length of the minor arc AB and the radius r are known. Which of the following can be determined?

     I. measure of angle α

     II. area of the shaded region

     III. perimeter of the shaded region

(A) I only   (B) I and II only   (C) I and III only

(D) II and III only   (E) I, II and III

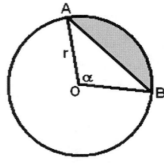

Figure 6

GO ON TO THE NEXT PAGE ▶▶▶

Model Test 1

45. $\displaystyle\sum_{k=1}^{99} \log\left(\frac{k}{k+1}\right) = ?$

(A) -100   (B) -2   (C) -1   (D) 2   (E) 100

46. The polar coordinates (r, θ) of the point (-2, 2) are

(A) $\left(-2\sqrt{2}, \frac{\pi}{4}\right)$ (B) $\left(2\sqrt{2}, -\frac{3\pi}{4}\right)$ (C) $\left(2\sqrt{2}, \frac{3\pi}{4}\right)$ (D) $\left(2, \frac{\pi}{2}\right)$ (E) $\left(8, \frac{3\pi}{4}\right)$

47. Defne has 2 pink, 2 green, 2 yellow, 2 blue, 2 white, and 2 purple socks in her sock drawer. She removes one sock at a time at random without replacement; at least how many socks must she remove in order to make sure that she has a pair of socks of the same color?

(A) five   (B) six   (C) seven   (D) eight   (E) more than eight

48. A plane intersects the cube given in figure 7 and satisfies the following three conditions:

    i. It passes through A

    ii. It is perpendicular to the base of the cube.

    iii. It divides the cube into two identical solids.

How many such planes are there?

(A) 2   (B) 4   (C) 6   (D) 8   (E) more than 8

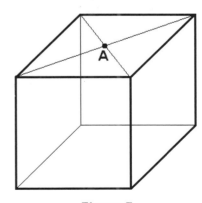

Figure 7

GO ON TO THE NEXT PAGE ▶

Model Test 1

49. The function f is defined as $f(x) = f(x-1) \cdot f(x-2)$. If $f(1) = 1$ and $f(2) = 2$ then $f(9) = ?$

(A) 13   (B) 21   (C) 2^{13} (D) 2^{21} (E) 2^{34}

50. What is the determinant of the coefficient matrix of the system of equations shown?

$$2x + 4y - z = 11$$
$$x - 2y + 5z = -3$$
$$x + y = 8$$

(A) -7   (B) -3   (C) 6   (D) 7   (E) 14

S T O P

END OF TEST

(Answers on page 199 – Solutions on page 203)

Model Test 2

Test Duration: 60 Minutes

Directions: For each of the following problems, decide which is the **best** of the choices given. If the exact numerical value is not one of the choices, select the choice that best approximates this value. Then fill in the corresponding oval on the answer sheet.

Notes:

- A calculator will be necessary for answering some (but not all) of the questions in this test. For each question you will have to decide whether or not you should use a calculator. The calculator you use must be at least a scientific calculator; programmable calculators and calculators that can display graphs are permitted.

- For some questions in this test you may have to decide whether your calculator should be in the radian mode or the degree mode.

- Figures that accompany problems in this test are intended to provide information useful in solving the problems. They are drawn as accurately as possible **except** when it is stated in a specific problem that its figure is not drawn to scale.

- All figures lie in a plane unless otherwise indicated.

- Unless otherwise specified, the domain of any function **f** is assumed to be the set of all real numbers **x** for which **f(x)** is a real number.

Reference Information: The following information is for your reference in answering some of the questions in this test.

- Volume of a right circular cone with radius **r** and height **h**: $V = \frac{1}{3}\pi r^2 h$

- Lateral area of a right circular cone with circumference of the base **c** and slant height **l**: $S = \frac{1}{2}cl$

- Volume of a sphere with radius **r**: $V = \frac{4}{3}\pi r^3$

- Surface area of sphere with radius **r**: $S = 4\pi r^2$

- Volume of a pyramid with base area **B** and height **h**: $V = \frac{1}{3}Bh$

1. If $x = \sqrt{9}$ and $y^2 = 16$ then $x + y$ can be

(A) − 1 or 7　　　(B) ±1 or ±7　　　(C) 1 or − 7　　　(D) − 1 or − 7　　　(E) 1 or 7

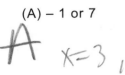

$x = 3, -3$　　$y = 4, -4$　　$7, -7$　　1.1052

2. The number given by 1.00001^{10000} is closest to

(A) e　　　(B) e^{10}　　　(C) $e^{0.1}$　　　(D) $\frac{3}{e}$　　　(E) $e^{-0.1}$

GO ON TO THE NEXT PAGE ▶▶▶

(31)

Model Test 2

3. If $\dfrac{4}{w+1} = \dfrac{w-5}{w-3}$ then w can be

(A) 1 only   (B) 7 only   (C) 1 or 7   (D) –1 or 1   (E) –1 or 7

$w^2 - 4w - 5 = 4w - 12$

$w^2 - 8w + 7 = 0$

$w \to 7$

4. Given that a = 4·ln(t) and b = 3·sin(t). What is the value of b when a is 8?

(A) 0.38   (B) 0.39   (C) 2.39   (D) 2.68   (E) 2.86

5. It is given that p/2 is an odd integer and p/5 is an even integer. If p is positive then which of the following must be correct?

  I. np is an even integer for all positive integer values of n.

  II. p^n is an odd integer for all positive integer values of n.

  III. $\log(p^n) > 1$ for all positive integer values of n.

(A) I only   (B) II only   (C) I and III only   (D) II and III only   (E) I, II and III

$p = 10$
30

$n \log p > 1$

6. If f(x) = $\sqrt{(x-2)^2 + 3}$ for x≥2; then f⁻¹(f(3))=?

(A) 0   (B) 2   (C) 3   (D) ±2   (E) ±3

$f(3) = 2, -2$ f^{-1}. $x = \sqrt{(y-2)^2+3}$

$\sqrt{x^2-3}+2 = y$

7. A is the set of all divisors of the number 60. What is the sum of the elements in set A?

(A) 0   (B) 12   (C) 17   (D) 77   (E) 168

60 1   5   12
2 30   6   10
3 20
4 15

GO ON TO THE NEXT PAGE ▶▶▶

Model Test 2

8. If the digits of a certain two-digit positive integer are interchanged, the resulting integer is 45 more than the original integer. What is the greatest possible value of the original integer?

(A) 27   (B) 38   (C) 49   (D) 83   (E) 94

38

49

9. The graph of y = mx + 1 has no points in the third and fourth quadrants if

(A) m = 0   (B) m < 0   (C) m > 0   (D) m < -1   (E) -1 < m < 0

10. If $2^{2x}+2^{-2x} = 5$ then $\dfrac{1}{x}$ can be

(A) 0.80 only   (B) – 0.89 or 0.89   (C) – 1.13 or 1.13   (D) 1.31 only   (E) – 0.88 or 0.88

 $x = 1.1302$

11. If $3 + \sqrt{2}$ is one root of a quadratic equation given by $x^2 – Px + Q = 0$ where P and Q are integers then Q is

(A) 0   (B) 6   (C) 7   (D) – 6   (E) – 7

 $3+\sqrt2 , 3-\sqrt2$

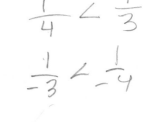

12. If $\dfrac{1}{p} < \dfrac{1}{q}$ and p < q then which of the following cannot be correct?

(A) p and q are both integers   (B) pq is positive
(C) p and q are both nonzero   (D) pq is negative
(E) p and q are both real

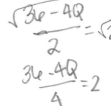

GO ON TO THE NEXT PAGE ▶▶▶

(33)

Model Test 2

13. If the point with coordinates (5, 7) is on the graph of f(x,y) = 0 which is symmetric with respect to the x – axis, which of the following points is also on the same graph?

(A) (7, -5)　　　　(B) (7, 5)　　　　(C) (-5, -7)　　　　(D) (-5, 7)　　　　(E) (5, -7)

14. If 3 – 2i is one root of a quadratic equation with real coefficients then the other root is

(A) – 3 + 2i　　　　(B) – 3 – 2i　　　　(C) 3 – 2i　　　　(D) 3 + 2i　　　　(E) 2 + 3i

15. If $\ln x = e^{-2x}$ then x can be

(A) 1.1138　　　　(B) 1.1318　　　　(C) 1.1381　　　　(D) 1.3181　　　　(E) 1.1813

16. $\cos^2(2x) - 1 = ?$

(A) $-\sin^2(2x)$　　　　(B) $\sin^2(2x)$　　　　(C) $\cos(2x)$　　　　(D) $2\sin^2(x)$　　　　(E) $2\sin^2(2x)$

17. Which of the following figures best describes the region that represents the set of all points (x,y) for which $|y| \le |x|$?

(A)	(B)	(C)	(D)	(E)

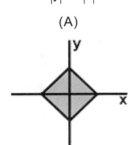 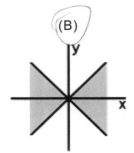 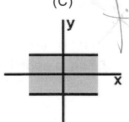 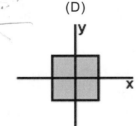 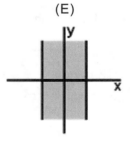

GO ON TO THE NEXT PAGE ▶▶▶

Model Test 2

18. In order to have at least two people born on the same day of week, we must have n people in a classroom. Which of the following is false?

(A) n can be odd

(B) n can be even

(C) n can be any positive multiple of 7

(D) n can be any number greater than 7

(E) n can be the sum of two distinct prime numbers greater than 2

19. If $\sqrt{4z-3}+\sqrt{8z+1}=8$ then the solution set for z

(A) is empty

(B) contains 2 rational numbers

(C) contains 1 rational number only

(D) contains 1 irrational number only

(E) contains 1 rational and 1 irrational number

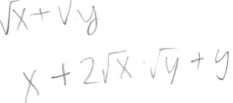

20. What is the range of $y = 4 \cdot \cos\theta$?

(A) $-1 \le y \le 1$ (B) $-1 < y < 1$ (C) $-\frac{1}{4} \le y \le \frac{1}{4}$ (D) $0 \le y \le 4$ (E) $-4 \le y \le 4$

21. In figure 1, rectangle OABC is rotated in the counterclockwise direction for θ degrees about point O, to get rectangle OA'B'C'. If the coordinates of B are (x, y) then what are the coordinates of point C'?

(A) $(x \cdot \cos\theta, x \cdot \sin\theta)$

(B) $(- y \cdot \sin\theta, y \cdot \cos\theta)$

(C) $(y \cdot \sin\theta, - y \cdot \cos\theta)$

(D) $(x \cdot \cos\theta - y \cdot \sin\theta, x \cdot \sin\theta + y \cdot \cos\theta)$

(E) $(x \cdot \cos\theta + y \cdot \sin\theta, x \cdot \sin\theta - y \cdot \cos\theta)$

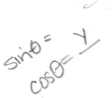

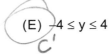

Figure 1

GO ON TO THE NEXT PAGE ▶▶▶

Model Test 2

22. If $(2x - 3)^2 - (4 - 5y)^2 = 6$ then the set of points (x,y) represents

(A) a circle   (B) an ellipse   (C) a parabola   (D) a hyperbola

(E) none of the above

23. If $g(x) = 2x - 1$ and $f(g(x)) = 3x + 4$ then $f^{-1}(x) = ?$

(A) $\dfrac{3x + 11}{2}$ (B) $\dfrac{2x - 11}{3}$ (C) $\dfrac{3x + 5}{2}$ (D) $\dfrac{2x - 5}{3}$

(E) none of the above

24. Domains of functions f(x), g(x) and f(g(x)) are three <u>distinct</u> sets A, B and C respectively. The function f(g(x)) is defined for all x in the set given by

(A) $A \cap B$ (B) $A \cup B$ (C) $B \cap C$ (D) $B \cup C$ (E) C

25. If $f(f(x)) = ax^2 + bx + c$ for some nonzero real values of a and b then f(x) is

(A) a constant function   (B) a linear function   (C) a quadratic function

(D) an increasing function   (E) not invertible

26. A three dimensional object has 6 vertices that are the centers of a cube having a volume of 216 cubic feet. If each face of this object is an isosceles triangle then what is the volume of the three dimensional object?

(A) 36   (B) 72   (C) 108   (D) 144   (E) 180

GO ON TO THE NEXT PAGE ►►►

Model Test 2

27. Graph of f(x) is given in figure 2. Which of the following is false?

(A) f(x) tends to zero as x becomes infinitely large or infinitely small.

(B) f(x) has one horizontal and two vertical asymptotes.

(C) f(x) has one local maximum at the point (0, −1).

(D) Domain of f(x) is all real numbers except ±3.

(E) Range of f(x) is all real numbers.

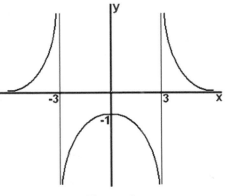

Figure 2

28. In figure 3; A, B and C are points on the circle whose center is at P. If B and C are fixed but A is a variable point on the major arc BC then the perimeter of triangle ABC is given by

(A) $2R\left(\sin\dfrac{\alpha}{2} + \cos\alpha + \cos\left(\dfrac{\alpha}{2} - x\right)\right)$

(B) $2R\left(\sin\dfrac{\alpha}{2} + \cos x + \cos\left(x - \dfrac{\alpha}{2}\right)\right)$

(C) $2R\left(\sin\dfrac{\alpha}{2} + \cos x + \cos(\alpha - x)\right)$

(D) $2R\left(\sin\dfrac{\alpha}{2} + \cos x + \cos\left(\dfrac{\alpha}{2} - x\right)\right)$

(E) $2R\left(\cos\dfrac{\alpha}{2} + c\sin x + \sin\left(\dfrac{\alpha}{2} - x\right)\right)$

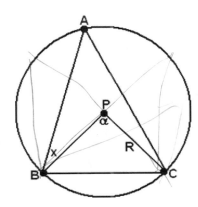

Figure 3

29. A mathematics teacher fixes two pegs on a table and attaches both ends of a loose chord on the pegs as shown in figure 4. The teacher then gets a piece of chalk and draws a curve on the table by stretching the chord with the chalk that she always keeps in contact with the table. The resulting curve on the table is part of

(A) a circle   (B) an ellipse   (C) a hyperbola   (D) a cardioid   (E) a parabola

Figure 4

GO ON TO THE NEXT PAGE ▶▶▶

Model Test 2

30. The function f(x) is given by f(x) = x·2^x. If $\frac{f(a+4)}{f(a+3)} = 6$ then a = ?

(A) −3.5　　　(B) −2.5　　　(C) 2.5　　　(D) 5.2　　　(E) 5.3

31. If the solution set of the equation given by cos(2x) = 1 + sin²y is not empty then how many solutions are there for x in the interval 0 ≤ x ≤ 2π?

(A) 1　　　(B) 2　　　(C) 3　　　(D) 4　　　(E) More than 4

32. If (sinx)(cscx) = 1 then x can be

(A) any real number　　　(B) any nonzero real number　　　(C) any odd multiple of $\frac{\pi}{2}$

(D) any real number that is not a positive multiple of π

(E) any real number that is not an integer multiple of π

33. How many solutions are there for the equation given by cos(3x) = $-x^2 + \frac{1}{2}$

(A) no solution　　　(B) 1　　　(C) 2　　　(D) 3　　　(E) 4

34. The graph of a parabola is given in figure 5. The coordinates of which of the following points are sufficient to determine the equation of the parabola?

　I. A and B

　II. A and V

　III. B and C

(A) I only　　(B) II only　　(C) III only　　(D) I or II only　　(E) II or III only

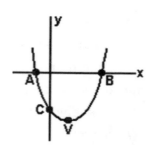

Figure 5

GO ON TO THE NEXT PAGE ▶▶▶

Model Test 2

35. The graph of f(x) is given in figure 6 and g(x) is defined in terms of f(x) as g(x) = f(x+1) − 1. How many zeros does g(x) have?

(A) 0   (B) 1   (C) 2

(D) 3   (E) 4

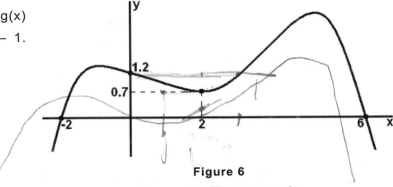

Figure 6

Figure not drawn to scale

36. What is the domain of the multivariable function given by $f(x,y) = \dfrac{\sqrt{x^2 + y^2 - 1}}{|x| + |y| + 1}$ in the x − y plane?

(A) All points on the Cartesian plane   (B) All points on and outside a circle

(C) All points on and inside a square   (D) All points on and inside a circle

(E) All points on and outside a square

37. If cosx = p and 90° < x < 180°, tanx=

(A) $\dfrac{\sqrt{1-p^2}}{p}$ (B) $\dfrac{p}{\sqrt{1-p^2}}$ (C) $\dfrac{p}{\sqrt{1+p^2}}$ (D) $-\dfrac{p}{\sqrt{1-p^2}}$ (E) $-\dfrac{\sqrt{1-p^2}}{p}$

38. If the measure of angle AOB in figure 7 is given as θ degrees then which of the following is correct?

(A) 90° < θ < 95°   (B) 95° < θ < 100°

(C) 100° < θ < 105°   (D) 105° < θ < 110°

(E) 110° < θ < 115°

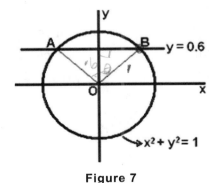

Figure 7

Figure not drawn to scale

GO ON TO THE NEXT PAGE ▶▶▶

Model Test 2

39. Two voyagers Nuran and Nazan are traveling from Antalya to Bodrum, covering a distance of 420 miles with the speeds of 105 and 100 mph respectively. Nuran starts at midnight and Nazan starts exactly 1 hour after Nuran makes it to Bodrum. When does Nazan arrive in Bodrum?

(A) 4:12 AM   (B) 4:20 AM   (C) 9:12 AM   (D) 9:12 PM   (E) 9:20 AM

40. A unitary number is one such that when its digits are added, when the digits of the resulting number are added, and so on, the final result is 1. For example, 8767 is a unitary number as $8 + 7 + 6 + 7 = 28$; $2 + 8 = 10$; and $1 + 0 = 1$. For a sequence, the first term is 1 and each term after the first term is the next greater unitary number. How many terms in this sequence are less than 200?

(A) 21   (B) 22   (C) 23   (D) 24   (E) 25

41. A rectangular puzzle consists of 7 rows and each row has 5 square pieces each. If a square piece is selected at random what is the probability that it has exactly 5 neighbors?

(A) $\dfrac{4}{7}$ (B) $\dfrac{7}{4}$ (C) $\dfrac{16}{35}$ (D) $\dfrac{35}{16}$

(E) None of the above

42. A ball is thrown horizontally from the top of a building 80 meters high and t seconds later; its position (x, y) is given by the parametric equations $x = 30t$ and $y = 80 - 5t^2$ (please note that initial position of the ball is given to be (0, 80)). How far in meters from the base of the tower does the ball land?

(A) 60   (B) 120   (C) 240   (D) 360   (E) 480

GO ON TO THE NEXT PAGE ▶▶▶

Model Test 2

43. The graph of f(x) is given in figure 8; $\frac{1}{f(x)}$ equals zero

when x is

(A) 1.8   (B) 2   (C) 3   (D) 4

(E) positive or negative infinity

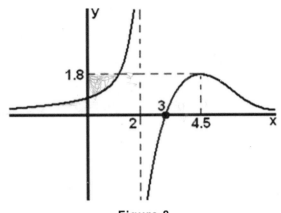

Figure 8
Figure not drawn to scale

44. On R^2 $2^x \triangle 3^y = 2x + 3y + 4$ is given; $8 \triangle 9 = ?$

(A) 16   (B) 17   (C) 36   (D) 46   (E) 47

45. Which of the following cannot be the range of a function whose domain is {-1, 1}?

(A) {0}   (B) {0, 1}   (C) {1, 2}   (D) {-1, 1}   (E) {-1, 0, 1}

46. If 3 − 2i and 3 + 2i are the roots of a quadratic equation given by $ax^2 + bx + c = 0$ where $i = \sqrt{-1}$
then which of the following is not correct?

(A) a is nonzero

(B) discriminant is negative

(C) a, b and c are all real or all complex

(D) a and c are both real whereas b can be real or complex

(E) $\frac{b}{a}$ and $\frac{c}{a}$ are both real

47. The solution set of $|x|^3 + 2|x|^2 - 3|x| = 0$ has how many elements?

(A) 0   (B) 1   (C) 2   (D) 3   (E) more than 3

GO ON TO THE NEXT PAGE ▶▶▶

Model Test 2

48. If in figure 9, AB = 4 inch and DC = 7 inch then what is the area of the shaded region rounded to the nearest tenth of a square inch?

   (A) 4.1   (B) 4.7   (C) 4.8   (D) 7.4   (E) 8.4

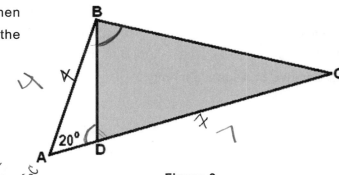

Figure 9

Figure not drawn to scale

49. In figure 10, the measure of θ is 0.03 radians and the base of the tree is 1.5 miles away from the nearest wall of the building. How high is the building to the nearest yard?

   (A) 79   (B) 84   (C) 94
   (D) 237   (E) 252

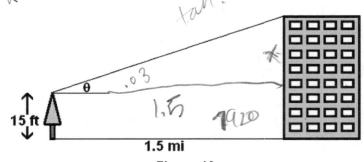

Figure 10

Figure not drawn to scale

50. What is the period of the function given by f(x) = |tan(3x)|

   (A) 3π   (B) 2π   (C) π   (D) π/2   (E) π/3

S T O P

END OF TEST

(Answers on page 199 – Solutions on page 206)

Model Test 3

Test Duration: 60 Minutes

Directions: For each of the following problems, decide which is the **best** of the choices given. If the exact numerical value is not one of the choices, select the choice that best approximates this value. Then fill in the corresponding oval on the answer sheet.

Notes:

- A calculator will be necessary for answering some (but not all) of the questions in this test. For each question you will have to decide whether or not you should use a calculator. The calculator you use must be at least a scientific calculator; programmable calculators and calculators that can display graphs are permitted.

- For some questions in this test you may have to decide whether your calculator should be in the radian mode or the degree mode.

- Figures that accompany problems in this test are intended to provide information useful in solving the problems. They are drawn as accurately as possible **except** when it is stated in a specific problem that its figure is not drawn to scale.

- All figures lie in a plane unless otherwise indicated.

- Unless otherwise specified, the domain of any function **f** is assumed to be the set of all real numbers **x** for which **f(x)** is a real number.

Reference Information: The following information is for your reference in answering some of the questions in this test.

- Volume of a right circular cone with radius **r** and height **h**: $V = \frac{1}{3}\pi r^2 h$

- Lateral area of a right circular cone with circumference of the base **c** and slant height **l**: $S = \frac{1}{2}cl$

- Volume of a sphere with radius **r**: $V = \frac{4}{3}\pi r^3$

- Surface area of sphere with radius **r**: $S = 4\pi r^2$

- Volume of a pyramid with base area **B** and height **h**: $V = \frac{1}{3}Bh$

1. What power of 34 equals 14 increased by 34%?

(A) 0.813   (B) 0.831   (C) 0.931   (D) 1.331   (E) 1.39

$$34^{x} = 18.76$$

2. If $\sqrt[3]{x-4} = 5$ then x=?

(A) -2.29   (B) 5.71   (C) 121   (D) 125   (E) 129

$$x - 4 = 125$$
$$x = 129$$

GO ON TO THE NEXT PAGE ▶▶▶

Model Test 3

3. It is given that f(g(x))=x and g(x)=2x+1; f(x)=?

(A) $\dfrac{x-1}{2}$ (B) $\dfrac{1-x}{2}$ (C) $\dfrac{x+1}{2}$ (D) $\dfrac{2}{x-1}$ (E) $\dfrac{1}{2x+1}$

$$f(2x+1) = X$$

4. Given that b=37(1.37)^a; c=$(3\pi)^{0.25}\left(\dfrac{b}{2}\right)^3$; and d=(2c-b)$^{0.5}$. If a=0.2 then d=?

(A) 39 (B) 163 (C) 164 (D) 26761 (E) 13400

$$b = 1.064...$$
$$c = .264...$$

5. If f(x) is a linear function where f(2) = 5 and f(-1)=2 then f(-3) = ?

(A) -2 (B) -1 (C) 0 (D) 1 (E) 2

6. The average of 10 numbers is 80 where the average of half of the numbers is 40. What is the average of the remaining half?

(A) 40 (B) 60 (C) 80 (D) 120

(E) None of the above

$$\frac{x}{10} = 80 \qquad \frac{y}{5} = 40$$
$$x = 800 \qquad y = 200$$
$$\frac{600}{5} = 120$$

7. f(x)=x^2-5; f(a)=9; a=?

(A) -14 (B) -3 (C) 3 (D) 3.74 (E) 14

$$a^2 - 5 = 9$$
$$a^2 = 14$$
$$a = \sqrt{14}$$

GO ON TO THE NEXT PAGE ▶▶▶

8. What is the distance between the points given by (4, 5, -2) and (1, -2, -6)?

(A) 8.6   (B) 9.05   (C) 9.06   (D) 9.89   (E) 9.90

$\frac{-7}{-3} = 2.33...$

-4

9. A group of students organize a trip so that the cost per person is $12.50 when a minimum of 20 people attend the trip. However when more people decide to join the trip, the total cost will be evenly divided among all people giving a lower cost per person. If n more people decide to attend the trip what will be the new cost per person in terms of n?

(A) $\dfrac{12.50 \cdot (20 - n)}{20 + n}$ (B) $\dfrac{12.50 \cdot (20 - n)}{20}$ (C) $\dfrac{12.50 \cdot (20 + n)}{20 - n}$ (D) $\dfrac{12.50 \cdot 20}{20 + n}$ (E) $\dfrac{12.50 \cdot (20 - n)}{n}$

10. If $x^{51} + 3x^{50} = k^2$ where k is an integer, then x can be

(A) 22   (B) 24   (C) 26   (D) 28   (E) 30

$X x^{50} + 3 x^{50}$

$x^{50}(3+x) = k^2$

$x^{25}\sqrt{x+3} = k$

11. First two terms a_1 and a_2 of an arithmetic sequence are 5 and 7 respectively. What is the first term in this sequence that exceeds 2000?

(A) a_{998} (B) a_{999} (C) a_{1000} (D) a_{1001} (E) a_{1999}

$5 + 7 + ...$

$a_n = a_1 + (n-1)d$

$a_{1000} = 5 + (999)2 = 2003$

$a_{998} = 5 + (997)2 = 1999$

12. If sin(x)=a and csc(x)=b then what is b in terms of a?

(A) $-a$ (B) $-\dfrac{1}{a}$ (C) $\dfrac{-1}{\sqrt{1-a^2}}$ (D) $\dfrac{1}{a}$ (E) $\dfrac{1}{\sqrt{1-a^2}}$

$\sin = a$

$\dfrac{1}{\sin} = b$ $b = \dfrac{1}{a}$

GO ON TO THE NEXT PAGE ▶▶▶

Model Test 3

13. Which of the following functions decreases as x takes on values from $\frac{\pi}{3}$ to $\frac{\pi}{2}$?

(A) -1/sin(x)   (B) 1/cos(x)   (C) sin(x)   (D) cos(x)   (E) tan(x)

14. If sin(x)=35° then x cannot be

(A) 8.768   (B) 6.940   (C) 2.484   (D) 0.657   (E) 0.574

15. If a·b=2c-1 then which of the following must be correct given that a, b, and c are all integers?

(A) a-b is odd   (B) a and b are both even   (C) exactly one of a and b is odd

(D) a+b is odd   (E) a and b are both odd

16. If both of the points A(0,-5) and B(5,0) are on the graph of f(x), then f(x) can be defined as

I. f(x) = |x – 5|

II. f(x) = x² – 4x – 5

III. f(x) = -5cos(πx)

(A) I only   (B) II only   (C) II or III only   (D) I or II only   (E) I, II or III

17. If the sum of a set of positive integers can be found by multiplying the number of terms in this set by the median, which of the following must be correct regarding this set of integers?

(A) The average of the least and greatest integers in this set equals the median.

(B) The average of the least and greatest integers in this set equals the mean.

(C) There is an even number of integers in this set.

(D) This set consists of consecutive integers.

(E) The median and the mean are equal.

GO ON TO THE NEXT PAGE ▶▶▶

Model Test 3

18. If x is positive and y is negative then x − y can be

   I. positive

   II. negative

   III. zero

(A) I only   (B) II only   (C) I or III only   (D) II or III only   (E) I, II or III

19. What is the domain of the function given by $f(x) = \dfrac{3x}{\sqrt{x^2-9}}$?

(A) $-3 \le x \le 3$ (B) $x < -3$ and $x > 3$ (C) $-3 < x < 3$ (D) $x < -3$ or $x > 3$

(E) None of the above

20. Which of the following points given in figure 1 satisfy the relations given by $y \le 0$ and $x \le y \le -x$?

(A) A   (B) B   (C) C   (D) D   (E) E

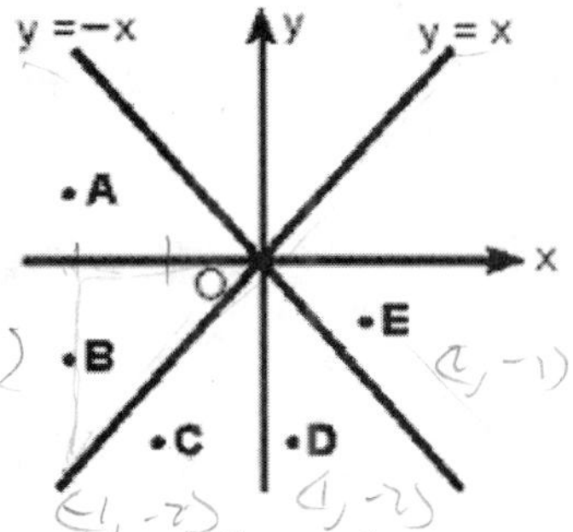

Figure 1

21. If the quantity given by $\dfrac{3m-1+a}{5m+6}$ represents a constant for all real values of m, then what is the value of a?

(A) -6/5   (B) 1   (C) 3/5   (D) 23/5   (E) It cannot be determined from the information given

22. Rectangle ABCD has its vertices located at (-1, -1), (5, -1), (5, 3), and (-1, 3). A line m divides rectangle ABCD into two congruent parts. If the equation of line m is $y = x + c$ then c is

(A) 5   (B) 3   (C) 1   (D) − 1   (E) − 3

GO ON TO THE NEXT PAGE ▶▶▶

Model Test 3

23. What is the amplitude of the function given by $f(x) = 4 \cdot \sin(3x) \cdot \cos(3x)$?

(A) -4 (B) -2 (C) 2   (D) 3   (E) 4

24. First two terms a_1 and a_2 of a geometric sequence are 5 and 7 respectively. What is the sum of the first 30 terms in this sequence rounded to the nearest integer?

(A) 216.072   (B) 302.505   (C) 423.513   (D) 216,072   (E) 302,505

$1\frac{2}{5} = \frac{7}{5}$

$$S_{30} = \frac{5 \left(1 - \frac{7}{5}\right)^{29}}{1 - \frac{7}{5}} =$$

25. If the quadratic function given by $f(x) = ax^2 + (b + c) \cdot x + (d + e) = 0$ intersects the x axis at two distinct points, then which of the following quantities must be nonzero?

(A) e   (B) d   (C) c   (D) b   (E) a

Parabola

26. There are totally 24 marbles in a bag of which 7 are blue, 8 are red and 9 are green. If two marbles are selected at random from the bag without replacement, what is the probability that they will be of different colors?

(A)	(B)	(C)	(D)	(E)
$\dfrac{7 \cdot 8 + 7 \cdot 9 + 8 \cdot 9}{48^2}$	$\dfrac{7 \cdot 8 + 7 \cdot 9 + 8 \cdot 9}{48 \cdot 47}$	$\dfrac{7 \cdot 8 + 7 \cdot 9 + 8 \cdot 9}{24^2}$	$\dfrac{7 \cdot 8 + 7 \cdot 9 + 8 \cdot 9}{24 \cdot 23}$	$\dfrac{2 \cdot (7 \cdot 8 + 7 \cdot 9 + 8 \cdot 9)}{24 \cdot 23}$

$24C_2$

27. If both the points (2, 5) and (-1, 2) are on the graph of $f(x) = \log_a(x+b)$ then b = ?

(A) 2.99   (B) 2.88   (C) 2.89   (D) -2.99   (E) -2.88

$a^5 - a^2 - 3 = 0$

$a = 1.373$

$a^5 = 2 + b$

$a^2 = -1 + b$

GO ON TO THE NEXT PAGE ▶▶▶

28. Graph of f(x) is given in figure 2. What is the graph of f(x+1)+1?

(A)

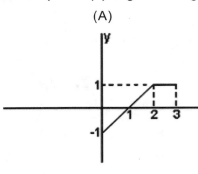

(B)

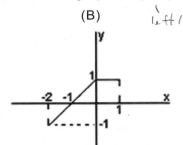

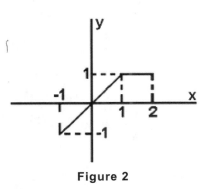

Figure 2

(D)

(E)

(C)

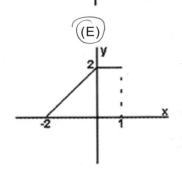

29. If f(x) satisfies the relation given by $f\left(\dfrac{a}{b}\right) = f(a) - f(b)$, then f(x) can be

(A) 3^x (B) logx   (C) sinx   (D) tanx   (E) x

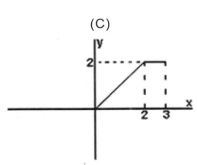

30. Given in figure 3 is a rectangular box whose dimensions are given in feet. If the measure of angle ∠DHB is 56° then what is the volume of the box rounded to the nearest cubic feet?

(A) 116   (B) 117   (C) 256   (D) 257

(E) There is not enough information to find the volume of this box.

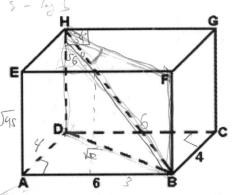

Figure 3

Figure not drawn to scale

GO ON TO THE NEXT PAGE ▶▶▶

(49)

Model Test 3

$$\begin{bmatrix} 1 & 2 & -3 \\ 2 & 1 & 0 \\ -1 & 1 & 1 \end{bmatrix} \begin{bmatrix} x \\ y \\ z \end{bmatrix} = \begin{bmatrix} 3 \\ 4 \\ -3 \end{bmatrix}$$

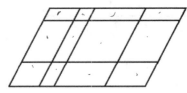

31. Which of the following is the system of equations represented by the matrix equation given above?

(A)	(B)	(C)	(D)	(E)
$x + 2y - z = 3$	$x + 2y - 3z = 3$	$x + 2y - 3z = 3$	$x + 2y - 3z = 3$	$x + 2y - 3z = -3$
$2x + y + z = 4$	$2x + y = 4$	$2x - y = 4$	$2x + y = 4$	$2x + y = 4$
$-3x + z = -3$	$-x + y + z = -3$	$-x + y + z = -3$	$x + y + z = -3$	$-x + y + z = 3$

32. Given in figure 4 are 5 parallel lines intersected by 4 other parallel lines. How many unique parallelograms are there in figure 4?

(A) 10   (B) 16   (C) 45   (D) 60

(E) None of the above

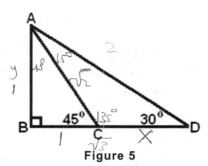

Figure 4

33. Given in figure 5 above, AB = y and CD = x. What is y in terms of x?

(A) $x(\sqrt{3} - 1)$ (B) $x(\sqrt{3} + 1)$ (C) $\dfrac{x\sqrt{3}}{\sqrt{3} - 1}$ (D) $\dfrac{x}{\sqrt{3} - 1}$ (E) $\dfrac{x}{\sqrt{3} + 1}$

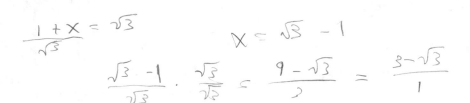

Figure 5

Figure not drawn to scale

34. If $4^x - x > 3$ then what is the solution set for x?

(A) x < -2.984 or x > 1   (B) -2.984 < x < 1   (C) x > 1   (D) x ≤ -2.984 or x ≥ 1   (E) x < -2.984

GO ON TO THE NEXT PAGE ▶▶▶

Model Test 3

35. How many 5 digit integers do not contain successive digits that are the same?

(A) 10^5 (B) 9^5 (C) $9 \cdot 10^4$ (D) $9^2 \cdot 8 \cdot 7 \cdot 6$ (E) $9 \cdot 8 \cdot 7 \cdot 6 \cdot 5$

36. Perimeter of ∆ABC is 11 cm and two of its sides are 3 cm and 4 cm in length. What is the smallest angle in this triangle rounded to the nearest degree?

(A) 22°   (B) 37°   (C) 44°   (D) 53°   (E) 68°

37. The cone given in figure 6 is cut into two parts with equal volumes by a slice made parallel to the base. What is the ratio of the area of resulting circle with center B to that of the circle with center C?

(A) $\dfrac{1}{2}$ (B) $\dfrac{1}{\sqrt{2}}$ (C) $\dfrac{1}{\sqrt[3]{2}}$ (D) $\dfrac{1}{\sqrt[3]{4}}$

(E) None of the above

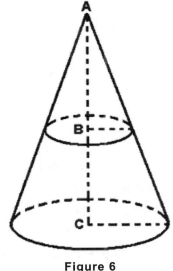

Figure 6
Figure not drawn to scale

38. Given in figure 7, POQ is an isosceles triangle where side PQ is tangent to the circle O at the point C. If length of chord AB (not shown) is 2x and POQ measures 2θ degrees, then what is the length of segment PQ in terms of x and θ?

(A) $\dfrac{x}{\sin\theta}$ (B) $\dfrac{2x}{\sin\theta}$ (C) $\dfrac{x}{2 \cdot \sin\theta}$ (D) $\dfrac{x}{\cos\theta}$ (E) $\dfrac{2x}{\cos\theta}$

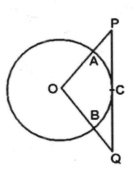

Figure 7

GO ON TO THE NEXT PAGE ▶▶▶

Model Test 3

39. What is the frequency of the function given by f(x) = 2·sin(πx+π)?

(A) π (B) $\dfrac{1}{\pi}$ (C) $\dfrac{1}{2}$ (D) 1   (E) 2

40. Which of the following is the inverse of the statement given by "If x=3, then x^2=9"?

(A) x≠3 or x^2=9.   (B) If x^2 = 9, then x=3.   (C) If x ≠ 3, then x^2≠9.

 (D) If x^2≠9, then x≠3.   (E) None of the above

41. $x - \dfrac{1}{x} = 4 \Rightarrow x^2 + \dfrac{1}{x^2} = ?$

(A) 10   (B) 12   (C) 14   (D) 16   (E) 18

42. Which of the following is not a correct representation of the shaded region given in figure 8?

(A) 1 - |y| ≤ 1 - |x|
(B) (y − 1)(1 + y) ≥ (x − 1)(1 + x)
(C) (3 − x^2)(3 + x^2) ≤ (3 − y^2)(3 + y^2)
(D) (2x − 1)(2x + 1) ≤ (2y − 1)(2y + 1)
(E) (2 − x^2)(x^2 + 2) ≥ (2 − y^2)(y^2 + 2)

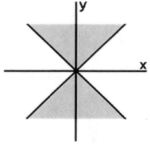

Figure 8

43. If sin(2x) = 1 + 4y^2 and 0 < x < 2π, then x = ?

(A) 0   (B) 1   (C) $\dfrac{\pi}{4}$ (D) $\dfrac{\pi}{2}$ (E) The given information is not enough to find x.

GO ON TO THE NEXT PAGE ▶▶▶

Model Test 3

44. The absolute value of a number x equals negative of itself. Which of the following gives all possible values of x?

(A) x > 0   (B) x < 0   (C) x ≤ 0   (D) x ≥ 0   (E) All real numbers

$$|x| = -x$$

45. Slopes of two intersecting lines are 3/7 and 15/7 respectively. Which of the following is the acute angle between these two lines rounded to the nearest degree?

(A) 42°   (B) 60°   (C) 120°   (D) 138°

(E) None of the above

46. What is the slope of the line given by $\begin{pmatrix} x \\ y \end{pmatrix} = \begin{pmatrix} 1 \\ 3 \end{pmatrix} + t\begin{pmatrix} 2 \\ 4 \end{pmatrix}$?

(A) 3   (B) 1/3   (C) 2   (D) ½

(E) None of the above

47. $4x^2 - 2kx + 1 = k - 2$

The equation given above has two real and distinct roots. Which of the following gives all possible values of k?

(A) -2 < k < 6   (B) -6 ≤ k ≤ 2   (C) k < -6 or k > 2   (D) k ≤ -6 or k ≥ 2   (E) -6 < k < 2

48. A square pyramid is inscribed in a hemisphere of radius 6 inches so that all vertices of the pyramid are on the hemisphere. What is the volume of the pyramid in cubic inches?

(A) 144   (B) 216   (C) 288   (D) 432   (E) 864

GO ON TO THE NEXT PAGE ▶▶▶

Model Test 3

49. $\lim_{x \to -1} \dfrac{x^2 - 1}{x + 1} = ?$

(A) -2   (B) -1   (C) 0   (D) 1   (E) Undefined

50. g(x) is a function defined in terms of another function f(x). If g(-x) = g(x) then g(x) can be defined for all f(x) as

    I. $g(x) = f^2(x)$

    II. $g(x) = f(-x) + f(x)$

    III. $g(x) = f(-|x|)$

(A) I only   (B) II only   (C) I or II only   (D) II or III only   (E) I, II or III

S T O P

END OF TEST

(Answers on page 199 – Solutions on page 209)

Model Test 4

Test Duration: 60 Minutes

Directions: For each of the following problems, decide which is the **best** of the choices given. If the exact numerical value is not one of the choices, select the choice that best approximates this value. Then fill in the corresponding oval on the answer sheet.

Notes:

- A calculator will be necessary for answering some (but not all) of the questions in this test. For each question you will have to decide whether or not you should use a calculator. The calculator you use must be at least a scientific calculator; programmable calculators and calculators that can display graphs are permitted.

- For some questions in this test you may have to decide whether your calculator should be in the radian mode or the degree mode.

- Figures that accompany problems in this test are intended to provide information useful in solving the problems. They are drawn as accurately as possible **except** when it is stated in a specific problem that its figure is not drawn to scale.

- All figures lie in a plane unless otherwise indicated.

- Unless otherwise specified, the domain of any function **f** is assumed to be the set of all real numbers **x** for which **f(x)** is a real number.

Reference Information: The following information is for your reference in answering some of the questions in this test.

- Volume of a right circular cone with radius **r** and height **h**: $V = \frac{1}{3}\pi r^2 h$

- Lateral area of a right circular cone with circumference of the base **c** and slant height **l**: $S = \frac{1}{2}cl$

- Volume of a sphere with radius **r**: $V = \frac{4}{3}\pi r^3$

- Surface area of sphere with radius **r**: $S = 4\pi r^2$

- Volume of a pyramid with base area **B** and height **h**: $V = \frac{1}{3}Bh$

1. If $x = e^{\ln 3} - \sqrt{3}$, then $\sqrt{x} =$

(A) 1.126   (B) 1.267   (C) 1.268   (D) 2.175   (E) 4.732

2. If $x + y = 13$, $x + z = 14$, and $y + z = 15$, then $x + y + z =$

(A) 16   (B) 17   (C) 18   (D) 21   (E) 42

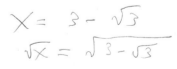

GO ON TO THE NEXT PAGE ▶▶▶

Model Test 4

$x = \dfrac{\log 5}{\log 5}$ $y = 3$

3. If $2^x = 5$ and $2^y = 8$ then $x - y$?

(A) – 0.68   (B) – 0.86   (C) 0.68   (D) 5.50   (E) 6.97

4. How many integers satisfy the inequality given by $|3x - 1| < 9$?

(A) Two   (B) Three   (C) Four   (D) Five   (E) More than five

$-1 \quad 0 \quad 1 \quad 2 \quad 3$

5. What is the degree equivalent of 2.5 radians?

(A) 0.795   (B) 0.796   (C) 134   (D) 143   (E) 314

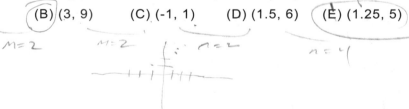

$\dfrac{\frac{5\pi}{2}}{x} = \dfrac{180}{\pi}$ $2.5 = \dfrac{180}{\pi} = 143.23$ $\dfrac{360°}{2\pi}$

6. All of the following points are on the same line except

(A) (2, 7)   (B) (3, 9)   (C) (-1, 1)   (D) (1.5, 6)   (E) (1.25, 5)

$M = 2$ $M = 2$ $n = 2$ $n = 4$

7. What is the product of the first seven consecutive multiples of 7?

(A) 196   (B) 823,543   (C) 84,707,280   (D) 4,150,656,720

(E) None of the above

7 14 21 28 35 42 49

8. All of the following expressions are equivalent. except

(A) –5(1–x)(x–4)   (B) –5(x–1)(4–x)   (C) –5(1–x)(4–x)   (D) 5(1–x)(4–x)   (E) 5(x–1)(x–4)

$x - 4 - x^2 + 4x$
$4x - x^2 - 4 - x$ $4 - x$

GO ON TO THE NEXT PAGE ▶▶▶

Model Test 4

9. How many of the elements in the set {e, ln(2), $\sqrt[4]{16}$, $\sqrt[3]{5}$, log1000, $\sqrt{3}$, ln(1), π} are irrational?

(A) 4   (B) 5   (C) 6   (D) 7   (E) all elements

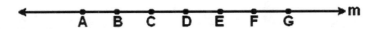

Figure 1

10. Two of the seven points on the line m given in figure 1 above are to be selected and connected to make a line segment. How many unique line segments can be constructed?

(A) 49   (B) 42   (C) 21   (D) 14   (E) 13

11. If $\dfrac{\sin^3 x + \sin x \cdot \cos^2 x}{\cos x} = 2.5$ when 0 < x < 90° then what is the measure of x rounded to the nearest degree?

(A) 21   (B) 22   (C) 32   (D) 68   (E) 86

12. An object is projected vertically upwards so that t seconds later its height in meters is given by h(t) = 100 + 20t − 5t². The object is at a height of 75 meters when

(A) t = 3   (B) t = 4   (C) t = 5   (D) t = 6   (E) t = 7

13. A periodic function given by $y = f(x)$ has a period of T. Which of the following is the period of

$y = \dfrac{1}{3} \cdot f(2x - 4) + 5$

(A) $\dfrac{T}{3}$ (B) 2T   (C) 4T   (D) $\dfrac{T}{2}$ (E) 5T

GO ON TO THE NEXT PAGE ▶▶▶

Model Test 4

14. The graph of an ellipse given by $\dfrac{x^2}{25} + \dfrac{y^2}{16} = 1$ is given in figure 2. If the foci are located at points P and Q, then PD + QD = ?

(A) 5   (B) 10   (C) 25   (D) 50

(E) It cannot be determined from the information given.

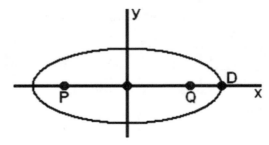

Figure 2

Figure not drawn to scale

15. The function $f(x) = \lfloor x \rfloor$ represents the greatest integer less than or equal to x and $g(x) = -\dfrac{10}{x}$; f(g(3)) + f(g(-6)) = ?

(A) -5   (B) -4   (C) -3   (D) 3   (E) 4

$$t_1 = 2$$
$$t_{n+1} = t_n + (2n + 1)$$

16. The recursive definition of a sequence t_n is given above. What is t_n in terms of n?

(A) $n^2 + 1$ (B) $(n - 1)^2 + 2$ (C) $(n + 1)^2 - 2$ (D) $(n + 1)^2 + 1$ (E) $(n - 1)^2 - 1$

17. If $f(x) = 2x^2 - 4x + 5$, which of the following is equal to f(-1) ?

(A) f(0)   (B) f(1)   (C) f(2)   (D) f(3)   (E) f(4)

18. The curve given by $y = x^3 - x^2$ is reflected across origin to get another curve given by $y = f(x)$. $f(x) = ?$

(A) $f(x) = \dfrac{1}{-x^3 + x^2}$ (B) $f(x) = -x^3 + x^2$ (C) $f(x) = -x^3 - x^2$ (D) $f(x) = x^3 + x^2$ (E) $f(x) = (x - 1)^3 + (x - 1)^2$

GO ON TO THE NEXT PAGE ▶▶▶

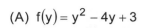
Model Test 4

19. The graph given in the figure 3 corresponds to the curve whose equation can be written as x = f(y); f(y) = ?

(A) $f(y) = y^2 - 4y + 3$

(B) $f(y) = \dfrac{y^2}{3} + \dfrac{4}{3}y + 1$

(C) $f(y) = \dfrac{y^2}{3} - \dfrac{4}{3}y + 3$

(D) $f(y) = \dfrac{1}{3}(y^2 - 4y + 3)$

(E) $f(x) = \dfrac{x^2}{3} - \dfrac{4}{3}x + 1$

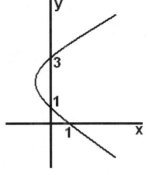

Figure 3

20. If f(x) = 2x − 1 and f(g(x)) = 3x + 4 then $g^{-1}(x)$=?

(A) $\dfrac{3x+5}{2}$

(B) $\dfrac{2x-5}{3}$

(C) $\dfrac{3x+11}{2}$

(D) $\dfrac{2x-11}{3}$

(E) None of the above

21. Which of the following figures can be rotated 180° about a horizontal axis to give the three dimensional object in figure 4?

(A)   (B)   (C)

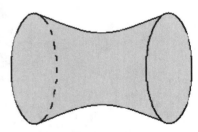

Figure 4

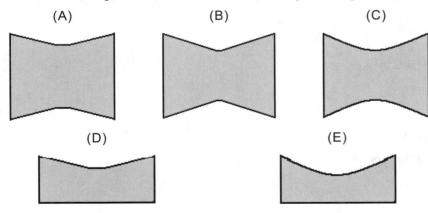

(D)   (E)

22. A die is biased so that when it is tossed, the probability of obtaining an odd number is 0.6. If this die is tossed 5 times, what is the probability of obtaining 3 odd numbers in succession?

(A) $0.6^3 \cdot 0.4^2$ (B) $0.6^2 \cdot 0.4^3$ (C) $0.6^3 \cdot 0.4^2 \cdot 3$ (D) $0.6^3 \cdot 0.4^2 \cdot C(5,3)$ (E) $0.6^3 \cdot 0.4^2 \cdot P(5,3)$

GO ON TO THE NEXT PAGE ▶▶▶

Model Test 4

23. In **RUSH** Academy every female teacher has half as many male colleagues as her female colleagues. If the number of male and female teachers in **RUSH** Academy are m and f then which of the following gives the relation between m and f correctly?

(A) f = 2m   (B) 2f = m   (C) 2f – 1 = m   (D) f – 1 = 2m   (E) 2(f – 1) = m

24. What is the amplitude of the function given by f(x) = – 3·sinx + 4·cosx + 1?

(A) – 5   (B) – 3   (C) 3   (D) 4   (E) 5

25. If x is an integer between the interval $-2 < x < 4$ and y is an integer between the interval $-3 \le y < 0$ then what is the maximum value of $(x + y)^2 = ?$

(A) 0   (B) 1   (C) 4   (D) 9   (E) 16

26. The expression $\left(1 + 2 + 2^2\right)\left(1 + 3 + 3^2 + 3^3\right)(1 + 5)$ gives

(A) The sum of all distinct divisors of 540

(B) The product of all distinct divisors of 540

(C) The sum of all distinct positive divisors of 540

(D) The product of all positive distinct divisors of 540

(E) None of the above

27. The function f(x) = 2·cos(x) is to be shifted $\frac{\pi}{2}$ units leftward to give g(x); g(x) = ?

(A) π·cos(x)   (B) 2π·cos(x)   (C) – 2·sin(x)   (D) 2·sin(x)   (E) 2·cos(x) + $\frac{\pi}{2}$

GO ON TO THE NEXT PAGE ▶▶▶

Model Test 4

28. What is the vertex of the parabola given by $x = 4(y+5)^2 + 1$

(A) (-5, -1) (B) (1, 5) (C) (5, 1) (D) (-5, 1) (E) (1, -5)

29. Which of the following are not sufficient to determine the equation of the parabola given in figure 5?

(A) A and C (C) B and C (D) B and D

(B) A, B and C (E) C, D and E

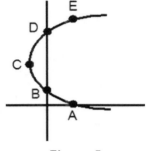

Figure 5

1. ABC is an acute triangle.

2. ABC is an isosceles triangle.

3. ABC is an obtuse triangle where B obtuse angle.

4. ABC is a right triangle where A is the right angle.

5. ABC is an obtuse triangle where A is the obtuse angle.

30. In a triangle ABC, $\sin A < \dfrac{a}{c}$ and $\cos A > \dfrac{b}{c}$. How many of the above statements are always false regarding triangle ABC?

(A) 1 (B) 2 (C) 3 (D) 4 (E) 5

I. $x_1 = x_2 = 1$ II. $x_1 = x_2 = \dfrac{2}{3}$ III. $x_1 = x_2 = 1 - \sqrt{3}$

IV. $x_1 = 1 + i; x_2 = 1 - I$ V. $x_1 = 1 - \sqrt{3}; x_2 = 1 + \sqrt{3}$

31. If x_1 and x_2 are the roots of a quadratic equation with integral coefficients then how many of the above are possible?

(A) 1 (B) 2 (C) 3 (D) 4 (E) all

GO ON TO THE NEXT PAGE ▶▶▶

Model Test 4

32. If $\sin\left(x+\dfrac{\pi}{2}\right) > \cos(x-\pi)$ where $0 < x < 2\pi$, then what is the solution set for x?

(A) $0 \le x < \dfrac{\pi}{2}$ or $\dfrac{3\pi}{2} < x \le 2\pi$ (B) $0 < x < 2$ (C) $\dfrac{\pi}{2} < x < \pi$

(D) $0 < x < \dfrac{\pi}{2}$ or $\dfrac{3\pi}{2} < x < 2\pi$ (E) $\dfrac{\pi}{2} < x < \dfrac{3\pi}{2}$

33. A cube has the same volume as the rectangular box shown in figure 6, whose dimensions are given in feet. If the rectangular box has a surface area of 376 square feet then what is one side of the cube in inches?

(A) 7.82   (B) 7.83   (C) 93.9   (D) 94.0   (E) 480

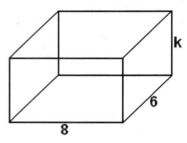

Figure 6

Figure not drawn to scale

34. If $\log_x 3 = 1.2$ and $\log_4 y = 5.6$, then $\log_x y = ?$

(A) 0.29   (B) 3.39   (C) 3.5   (D) 8.48   (E) 6.12

35. Graph of f(x) is given in figure 7. At which of the following intervals does f(x) increase throughout?

(A) x < 0

(B) x < -3

(C) x > 3

(D) -3 < x < 3

(E) x < -3 or x > 3

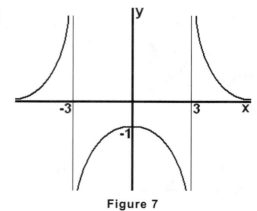

Figure 7

GO ON TO THE NEXT PAGE ▶▶▶

Model Test 4

36. The graphs of two functions f(x) and g(x) are plotted on the same coordinate plane as shown in figure 8. Which of the following is false?

(A) f(x) has no zeros whereas g(x) has more than two zeros

(B) f(x) and g(x) are both continuous functions

(C) f(x) − g(x) ≥ 0 for all x

(D) f(x)·g(x) intersects the x axis three times

(E) $\dfrac{f(x)}{g(x)}$ intersects the x axis more than once

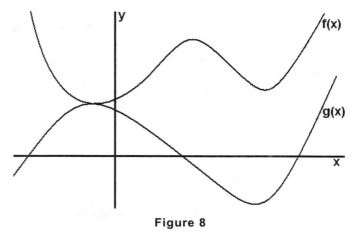

Figure 8

37. In figure 9 the coordinates of A and B are (-6, 5) and (2, 9). If points A, B and C are collinear and AB = 2BC then what are the coordinates of C?

(A) (4, 2)   (B) (4, 10)   (C) (6, 11)   (D) (5.5, 10.5)

(E) None of the above

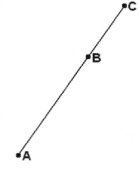

Figure 9

38. If the graph of f(x) is given in figure 10, which of the following is false?

(A)   f(x) has exactly two zeros

(B)   1/f(x) has exactly two zeros

(C)   range of f(x) is all real numbers

(D)   f(x) has one horizontal and two vertical asymptotes

(E)   when x increases without bound, f(x) increases without bound

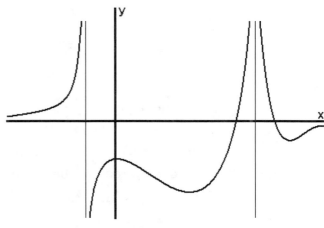

Figure 10

GO ON TO THE NEXT PAGE ▶▶▶

39. A curve given by $y = x^2 - 1$ is to be represented in a coordinate plane where the origin is translated to (1, -1) and each point on the curve stays fixed at its original position. The equation of the curve becomes

(A) $y = (x-1)^2 + 1$ (B) $y = (x-1)^2$ (C) $y = (x+1)^2 + 1$ (D) $y = (x+1)^2 - 1$ (E) $y = (x+1)^2$

40. If $f(x) = f(\pi - x)$ for all x in the domain, then which of the following could be f ?

(A) $f(x) = \sin(x)$ (B) $f(x) = \cos(x)$ (C) $f(x) = \tan(x)$ (D) $f(x) = \cot(x)$ (E) $f(x) = \sec(x)$

41. The graph of a polynomial function P(x) is given in figure 11 above. Which of the following is correct?

(A)   P(x) has a positive constant term.

(B)   P(x) has a negative leading coefficient.

(C)   P(x) has two local maxima and one local minimum.

(D)   Degree of P(x) can be any odd number greater than 3.

(E)   The function given by $f(x) = \dfrac{1}{P(x)}$ has more than 3 vertical

asymptotes.

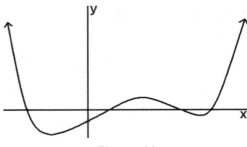

Figure 11

I. The sine function

II. A periodic function

III. A piecewise function

IV. The greatest integer function

V. The natural logarithm function

42. It is given that x and y are real numbers such that $0 < x < y < 1$. If $f(x) = f(y)$ then how many of the above can be used to define f(x)?

(A) 1   (B) 2   (C) 3   (D) 4   (E) 5

GO ON TO THE NEXT PAGE ▶▶▶

Model Test 4

43. Which of the following can be the equation of the curve given by |x + 4| + |y − 4| = 3?

(A)   (B)   (C)   (D)   (E)

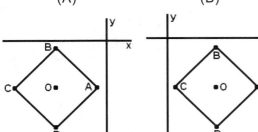 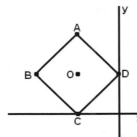 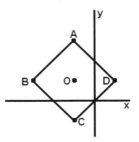 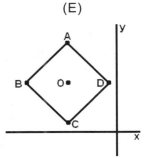

44. The author of this book says, "I usually do not sleep more than three hours a day but occasionally I sleep for 17 hours. By occasionally I mean once every 90 days." Based on this information what is the maximum number of hours the author may sleep in a 365 day period?

(A) 1151   (B) 1156   (C) 1163   (D) 1165   (E) 1180

45. For which of the following listings is the standard deviation the greatest?

(A) 22, 22, 22   (B) 22, 23, 24   (C) 22, 24, 26   (D) 15, 20, 25   (E) 11, 13, 15

46. The graph of $y = 3 \cdot \sin\left(x + \dfrac{\pi}{4}\right)$ is given in figure 12. If A is a point on this curve then p = ?

(A) 2.39   (B) 2.93   (C) 3.29   (D) 3.92   (E) 0.977

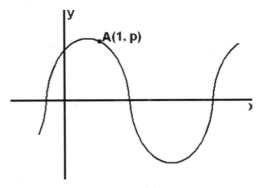

Figure 12

GO ON TO THE NEXT PAGE ►►►

Model Test 4

47. As x approaches 1, $\dfrac{x^3 - 1}{\ln x}$ approaches

(A) -3   (B) 0   (C) 1   (D) 3

(E) None of the above

48. For the matrices **P**, **Q**, **R** and **S** given in the table above, which of the following matrix multiplications cannot be carried out?

(A) **PQ**   (B) **RP**   (C) **QS**   (D) **SR**   (E) **PS**

Matrix	Dimensions
P	3×4
Q	4×5
R	2×3
S	5×2

$3 \times \boxed{4 \ 5}$

49. x and y are real numbers such that $0 < y < y^x < 1$. Which of the following includes all possible values of x?

(A) $x < -1$ (B) $-1 < x < 0$ (C) $x < 0$ (D) $0 < x < 1$ (E) $x > 1$

$y = \frac{1}{2}$

$\frac{1}{y^2}$ $0 < \frac{1}{2} < \frac{1}{2}^{\frac{1}{2}} < 1$

1. N = 0 and Q = 1
2. If Q < 0.000001 then go to 5 otherwise go to 3
3. Divide Q by 3 and increase N by 1 $Q = \frac{1}{3}$ $N = 1$
4. Go to 2
5. Print N $Q = \frac{1}{9}$ $N = 2$

                        $\frac{1}{27}$ 3

50. If the above instructions are carried out, what number will be printed?

(A) 11   (B) 12   (C) 13   (D) 14   (E) 15

S T O P

END OF TEST

(Answers on page 199 – Solutions on page 212)

Model Test 5

Test Duration: 60 Minutes

Directions: For each of the following problems, decide which is the **best** of the choices given. If the exact numerical value is not one of the choices, select the choice that best approximates this value. Then fill in the corresponding oval on the answer sheet.

Notes:

- A calculator will be necessary for answering some (but not all) of the questions in this test. For each question you will have to decide whether or not you should use a calculator. The calculator you use must be at least a scientific calculator; programmable calculators and calculators that can display graphs are permitted.

- For some questions in this test you may have to decide whether your calculator should be in the radian mode or the degree mode.

- Figures that accompany problems in this test are intended to provide information useful in solving the problems. They are drawn as accurately as possible **except** when it is stated in a specific problem that its figure is not drawn to scale.

- All figures lie in a plane unless otherwise indicated.

- Unless otherwise specified, the domain of any function **f** is assumed to be the set of all real numbers **x** for which **f(x)** is a real number.

Reference Information: The following information is for your reference in answering some of the questions in this test.

- Volume of a right circular cone with radius **r** and height **h**: $V = \frac{1}{3}\pi r^2 h$

- Lateral area of a right circular cone with circumference of the base **c** and slant height **l**: $S = \frac{1}{2}cl$

- Volume of a sphere with radius **r**: $V = \frac{4}{3}\pi r^3$

- Surface area of sphere with radius **r**: $S = 4\pi r^2$

- Volume of a pyramid with base area **B** and height **h**: $V = \frac{1}{3}Bh$

1. What is the value of $(1.208)^7$ correct to the nearest thousandth?

(A) 3.8   (B) 3.75   (C) 3.754   (D) 3.7538   (E) 3.75378

2. Which of the following is equivalent to $\sqrt{p} \cdot \sqrt[3]{p^2}$

(A) $\sqrt[7]{p^6}$ (B) $\sqrt[6]{p^5}$ (C) $p \cdot \sqrt[6]{p}$ (D) $p \cdot \sqrt[3]{p^2}$ (E) $p \cdot \sqrt[3]{p^4}$

GO ON TO THE NEXT PAGE ▶▶▶

Model Test 5

3. What is the amplitude of the function given by $f(x) = 3 \cdot \sin^2(2x+3\pi) + 1$?

(A) -3   (B) -1.5   (C) 1.5   (D) 3

(E) None of the above

4. A quadratic function is given by $f(x)=ax^2+bx+c$ where a, b, and c are all nonzero real numbers. If exactly two points are common to both $f(x)$ and $-f(x)$, then which of the following must be correct about $f(x)$?

(A) a is positive   (B) b + c is nonzero   (C) b is negative

(D) f(x) is nonnegative   (E) discriminant is positive

5. As x increases from 0 to 2π, sin(x) and cos(x) both decrease in

(A) Quadrant I   (B) Quadrant II   (C) Quadrant III

(D) Quadrant IV   (E) No quadrants

6. The graph of the equation $16y^2 - 9x^2 = 144$ represents

(A) a circle   (B) a hyperbola   (C) an ellipse   (D) a parabola   (E) a lemniscate

7. If one zero of a quadratic integral polynomial is $4 + 2i$, where $i^2 = -1$, the polynomial is

(A) $x^2 + 8x + 20$ (B) $x^2 - 8x + 20$ (C) $x^2 + 8x - 20$ (D) $x^2 - 8x - 20$

(E) None of the above

GO ON TO THE NEXT PAGE ▶▶▶

Model Test 5

8. If $-\sqrt{z-1} = (y+3)^2$ then y + z = ?

(A) -3   (B) -2   (C) 1   (D) 2   (E) 4

9. The locus of points at a distance of 3 inches from each of two perpendicular planes is

(A) 4 points   (B) 2 lines   (C) 4 lines   (D) 2 planes   (E) 4 planes

10. What is the smallest number greater than 1000 that gives the remainder of 3 when divided by 4, 5, and 6?

(A) 123   (B) 1003   (C) 1017   (D) 1020   (E) 1023

11. In which of the following intervals is the function given by $f(x) = x \cdot (x+5) \cdot (x-4)^2$ invertible?

(A) -7 < x < -2   (B) -5 < x < -1   (C) -1 < x < 2   (D) -2 < x < 1   (E) 1 < x < 3

12. If $\tan x = \dfrac{12}{5}$ and x is in the 3rd quadrant then sinx=?

(A) $\dfrac{12}{13}$ (B) $\dfrac{5}{13}$ (C) $\dfrac{5}{12}$ (D) $-\dfrac{5}{13}$ (E) $-\dfrac{12}{13}$

13. What is the solution of the system of inequalities given by $|2x+5| \leq 11$ and $2x+5 < 0$?

(A) x ≥ -8   (B) $x < -\dfrac{5}{2}$ (C) -8 ≤ x ≤ 3   (D) $-8 \leq x < -\dfrac{5}{2}$

(E) None of the above

GO ON TO THE NEXT PAGE ▶▶▶

14. Given that Cos(72°)=0.3090 and Cos(73°)=0.2924. If by interpolation the value of Cos(72.5°) is determined to be 0.3007, what is the assumption that has been made?

(A) Cos(x) is a constant function of x in the vicinity of x=72.5°

(B) Cos(x) is a linear function of x in the vicinity of x=72.5°

(C) Cos(x) is a quadratic function of x in the vicinity of x=72.5°

(D) Cos(x) is a cubic function of x in the vicinity of x=72.5°

(E) None of the above

15. Which of the following is logically equivalent to the statement given by p⇒q?

I. ~pVq

II. q⇒p

III. ~q⇒~p

(A) I only
(B) II only
(C) I, II and III
(D) I and III only
(E) II and III only

16. Two angles of a triangle measure 45° and 105°. What is the ratio of the longest side to the shortest side in this triangle?

(A) $\frac{(\sqrt{3}+1)\sqrt{2}}{2}$
(B) $\frac{(\sqrt{3}-1)\sqrt{2}}{4}$
(C) $\frac{(\sqrt{3}+1)}{2}$
(D) $\frac{(\sqrt{3}-1)}{4}$

(E) It cannot be determined from the given information.

17. The graph of r = 4·secθ represents

(A) a circle.
(B) an ellipse.
(C) a parabola.
(D) a hyperbola.
(E) a line.

GO ON TO THE NEXT PAGE ▶▶▶

Model Test 5

C = the set of complex numbers   Q = the set of rational numbers

N = the set of natural numbers   R = the set of real numbers

Z = the set of integers

18. Which of the following is a correct statement for the sets given above?

(A) $Z \subset N \subset Q \subset R \subset C$ (B) $C \subset N \subset Q \subset R \subset Z$ (C) $N \subset Z \subset Q \subset R \subset C$

(D) $C \subset R \subset Q \subset Z \subset N$ (E) $N \subset Z \subset R \subset Q \subset C$

19. If $a^x \cdot a^3 = a^4$ and $(a^y)^5 = a^{20}$ then x+y=?

(A) 1   (B) 4   (C) 5   (D) 16   (E) It cannot be determined from the information given

20. tan(x) + cot(x)=?

(A) sin(2x)   (B) $2 \cdot \sec(2x)$ (C) $2 \cdot \csc(2x)$ (D) $2 \cdot \sec(x) \cdot \csc(x)$

(E) none of the above

21. $\sin\left(\cot^{-1}\left(-\dfrac{3}{4}\right)\right) = ?$

(A) $\dfrac{4}{5}$ (B) $-\dfrac{4}{5}$ (C) $\dfrac{3}{5}$ (D) $-\dfrac{3}{5}$ (E) $\dfrac{4}{3}$

22. What is the area bounded by the closed curve represented by the equation given by $16x^2+9y^2-32x+36y=92$?

(A) 9π (B) 12π (C) 16π (D) 144π (E) It cannot be determined from the information given.

GO ON TO THE NEXT PAGE ▶▶▶

Model Test 5

$$x+2y+z=0$$
$$2x-y-z=0$$
$$-4x+Ey+2z=0$$

23. For what value of E does the above system of equations have a solution other than (0,0,0)?

(A) 2　　　　　(B) 4　　　　　(C) -4　　　　　(D) -8　　　　　(E) -2

24. If $x = 2 - e^{-t}$ and $y = 1 + e^{t}$ then what is the relation between x and y independent of t?

(A) $y - 1 = \dfrac{1}{x-2}$　　(B) $y - 1 = \dfrac{1}{x+2}$　　(C) $y + 1 = \dfrac{1}{x+2}$　　(D) $y + 1 = \dfrac{1}{x-2}$　　(E) $y - 1 = \dfrac{1}{2-x}$

1. start with f(x)
2. shift the resulting function in the previous step upwards for 1 unit
3. shift the resulting function in the previous step leftwards for 3 units
4. reflect the resulting function in the previous step with respect to the x axis
5. shrink the resulting function in the previous step in x by a factor of 0.5

25. Starting with f(x) we would like to end up with 1 – f(2x+6) at the end of the last step after performing all of the five steps given above not necessarily in the same order. What must be the correct order of the steps to be performed?

(A) **1 – 3 – 5 – 4 – 2**　　　(B) **1 – 2 – 5 – 3 – 4**　　　(C) **1 – 5 – 3 – 2 – 4**　　　(D) **1 – 5 – 3 – 4 – 2**

(E) There is an error in one of the steps therefore we cannot end up with 1 – f(2x+6) starting with f(x).

26. A solution is made by mixing three substances A, B and C in such a way that the ratio by weight of A to B is 5:8 and that of B to C is 6:11. If there are 48 pounds of substance B, what is the weight in ounces of the total mixture?

(A) 83　　　　　(B) 166　　　　　(C) 1328　　　　　(D) 2656

(E) None of the above

GO ON TO THE NEXT PAGE ▶▶▶

27. Graph of f(x) is given in figure 1 above. What will be the graph of -|f(-x)|?

(A)

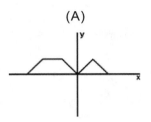

(B)

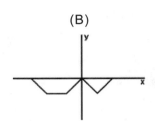

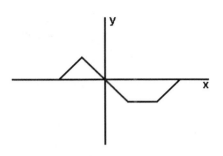

Figure 1

(C)

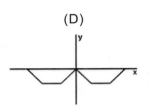

(D)

(E)

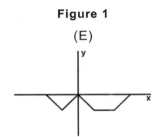

28. Each side of the regular pentagon ABCDE is 7 inches long. What is the area in square inches of ΔADC if the immediate neighbors of vertex A are the vertices B and E?

(A) 76   (B) 75   (C) 38   (D) 37   (E) It cannot be determined from the given information

29. What is the distance between points E and R whose coordinates on the oblique coordinate system given in figure 2 are (5, 4) and (-4, -6) respectively?

(A) 5.01   (B) 9.54   (C) 13.5   (D) 13.6

(E) none of the above

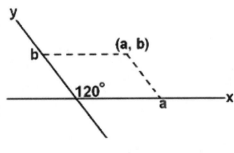

Figure 2

30. The graph of |3x| + |2y| = 6A where A is a positive real number consists of

(A) two intersecting lines   (B) two disjoint angles   (C) the sides of a square

(D) the sides of a rhombus   (E) the sides of a trapezoid

GO ON TO THE NEXT PAGE ►►►

Model Test 5

31. The portion of the surface, given by 3x + 4y + 5z = 18, that lies in the first octant forms a three dimensional object with the coordinate planes. What is the volume of this object?

(A) 10   (B) 16.2   (C) 32.4   (D) 48.6   (E) 97.2

32. What is the infinite product given by $5 \cdot 5^{1/2} \cdot 5^{1/4} \cdot 5^{1/8} \cdot 5^{1/16}\ldots$ equal to?

(A) 0.2   (B) 1   (C) 5   (D) 25   (E) 125

33. The equation given by $9x^2 - axy + 4y^2 = 16$ represents a pair of straight lines. The value of a can be

    I. 6   II. 12   III. -12

(A) I only   (B) II only   (C) III only   (D) I or II only   (E) II or III only

34. Which of the following is a solution of the equation given by $e^x = x+2$?

(A) -1.146   (B) -2.841   (C) 0.146   (D) 1.146   (E) 1.841

35. The shaded pentagon ABCDE given in figure 3 is rotated about line n for 360°. What is the volume of the solid that results?

(A) 58   (B) 32π   (C) 458π   (D) 490π

(E) None of the above

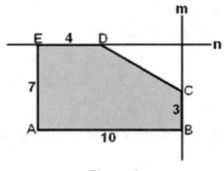

Figure 3

GO ON TO THE NEXT PAGE ▶▶▶

Model Test 5

36. What is the real polynomial of lowest degree such that two of its zeros are 3 and 2-i where i=$\sqrt{-1}$?

(A) $x^3-5x^2+11x-15$ (B) $x^3-5x^2+11x+15$ (C) $x^3-17x^2+7x-15$

(D) $x^3-7x^2+17x-15$ (E) $x^3+17x^2+7x+15$

37. The graph of x < -32 in a linear coordinate system is composed of

(A) a ray (B) a point (C) a region (D) a half line (E) a subspace

38. What is the solution set of the inequality given by $x^3 + 4 > 3x^2$?

(A) x < -1 (B) x > -2 and x ≠ 1 (C) x > -1 (D) -1 < x < 2 or x > 2 (E) x < -2

39. In figure 4 the sphere just fits in the cylinder. What is the volume outside the sphere and inside the cylinder in cubic inches if the radius of the sphere is R inches?

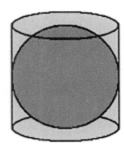

(A) $\frac{2}{3}\pi R^3$ (B) πR^3 (C) $\frac{4}{3}\pi R^3$ (D) $2\pi R^3$

(E) None of the above

Figure 4

40. Which of the following is correct about the triangles EBR such that in each triangle the measure of angle E is 27°; side e has a length of 33 inches and all other sides and angles can vary?

(A) Their areas are the same.

(B) They are similar triangles.

(C) They are congruent triangles.

(D) Their inscribed circles are congruent.

(E) Their circumscribed circles are congruent.

GO ON TO THE NEXT PAGE ▶▶▶

Model Test 5

41. If $f(x) = x^2 + 4x + 4$ the limit given by $\lim\limits_{h \to 0} \dfrac{f(x) - f(h)}{x - h}$ is

(A) positive infinity

(B) a linear functionof x

(C) undefined

(D) negative infinity

(E) a constant number

42. If $f(x) = \dfrac{x^2 - x - 2}{-2x^2 + 2x + 4}$ then for what values of x is $f(x) = \dfrac{-1}{2}$?

(A) All real numbers except for 1 and -2

(B) -1 or 2 only

(C) All real numbers except for -1 and 2

(D) 1 or -2 only

(E) All real numbers

43. $2 - \cfrac{1}{2 - \cfrac{1}{2 - \cfrac{1}{2 - \cfrac{1}{...}}}} = ?$

(A) 0   (B) 1   (C) 2   (D) 1/2   (E) -1/2

44. $f(x) = x^2 + x + 1$ and x takes integer values only. Which of the following must be correct about f(a) if a is a positive integer greater than 50 and the remainder is 1 when a is divided by 5?

(A) It is not a prime number.

(B) It is an odd number.

(C) It is a prime number.

(D) It is an even number.

(E) It is a multiple of 7.

GO ON TO THE NEXT PAGE ▶▶▶

45. The circle with radius of length 10 inches given in figure 5 is partitioned into two sectors and each sector is bent into a cone. If the minor and major sectors are bent into cones with volumes of V_1 and V_2 respectively, then $\dfrac{V_1}{V_2} = ?$

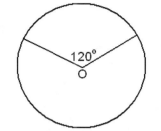

Figure 5

(A) $\dfrac{\sqrt{10}}{5}$

(B) $\dfrac{\sqrt{10}}{10}$

(C) $\dfrac{\sqrt{5}}{2}$

(D) $\dfrac{\sqrt{5}}{5}$

(E) None of the above

46. $\ln(\cos(2x)) = ?$

(A) $\ln(\cos x + \sin x) + \ln(\cos x - \sin x)$

(B) $2 \cdot \ln(\cos x) - 2 \cdot \ln(\sin x)$

(C) $\dfrac{\ln(\cos^2 x)}{\ln(\sin^2 x)}$

(D) $\ln(\cos x + 1) + \ln(\cos x - 1)$

(E) $\ln 2 + \ln(\cos x)$

47. Given that $|f(x)| = \left| f\left(\dfrac{1}{x}\right) \right|$ which of the following can be f(x)?

    I. $f(x) = x + \dfrac{1}{x}$
    II. $f(x) = x - \dfrac{1}{x}$
    III. $f(x) = 2x - \dfrac{1}{2x}$

(A) I only   (B) II only   (C) III only   (D) I or II only   (E) II or III only

48. What is the sum of the additive and multiplicative inverses of $1 + \sqrt{3}$?

(A) $\dfrac{3 + \sqrt{3}}{2}$

(B) $\dfrac{-3 - \sqrt{3}}{2}$

(C) $-1 - \sqrt{3}$

(D) $\dfrac{\sqrt{3} - 1}{2}$

(E) $\dfrac{3\sqrt{3} + 1}{2}$

GO ON TO THE NEXT PAGE ▶▶▶

Model Test 5

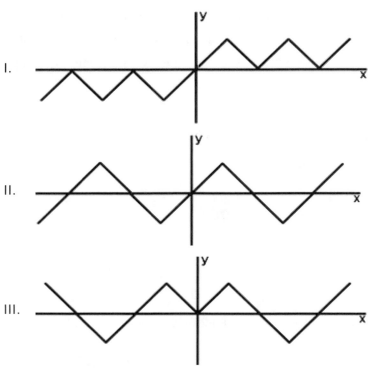

I.

II.

III.

49. A function f(x) satisfies the condition that |f(x)| is periodic whereas f(x) is not. Which of the above can be part of the graph of f(x)?

(A) I only   (B) II only   (C) III only   (D) I and III only   (E) I, II and III

50. Given that $f(x) = \dfrac{(x-1)}{(x+1)^2 \cdot (x+2)}$ which of the following is correct about f(x)?

(A) f(x) has three zeros and no horizontal or vertical asymptotes

(B) f(x) has one horizontal asymptote, two vertical asymptotes and no zeros.

(C) f(x) has one zero, one horizontal asymptote and two vertical asymptotes.

(D) f(x) has one zero, one horizontal asymptote and three vertical asymptotes.

(E) f(x) has one zero, three vertical asymptotes and no horizontal asymptotes.

S T O P

END OF TEST

(Answers on page 199 – Solutions on page 214)

Model Test 6

Test Duration: 60 Minutes

Directions: For each of the following problems, decide which is the **best** of the choices given. If the exact numerical value is not one of the choices, select the choice that best approximates this value. Then fill in the corresponding oval on the answer sheet.

Notes:

- A calculator will be necessary for answering some (but not all) of the questions in this test. For each question you will have to decide whether or not you should use a calculator. The calculator you use must be at least a scientific calculator; programmable calculators and calculators that can display graphs are permitted.

- For some questions in this test you may have to decide whether your calculator should be in the radian mode or the degree mode.

- Figures that accompany problems in this test are intended to provide information useful in solving the problems. They are drawn as accurately as possible **except** when it is stated in a specific problem that its figure is not drawn to scale.

- All figures lie in a plane unless otherwise indicated.

- Unless otherwise specified, the domain of any function **f** is assumed to be the set of all real numbers **x** for which **f(x)** is a real number.

Reference Information: The following information is for your reference in answering some of the questions in this test.

- Volume of a right circular cone with radius **r** and height **h**: $V = \frac{1}{3}\pi r^2 h$

- Lateral area of a right circular cone with circumference of the base **c** and slant height **l**: $S = \frac{1}{2}cl$

- Volume of a sphere with radius **r**: $V = \frac{4}{3}\pi r^3$

- Surface area of sphere with radius **r**: $S = 4\pi r^2$

- Volume of a pyramid with base area **B** and height **h**: $V = \frac{1}{3}Bh$

1. If $3m = e^2 + \pi + \sqrt{3}$ then $2m = ?$

(A) 4.08   (B) 4.09   (C) 8.17   (D) 8.18   (E) 12.26

2. If $e^x = 4 + \ln x$ then $x = ?$

(A) 1.45   (B) 1.46   (C) 1.47   (D) 1.48   (E) 1.49

GO ON TO THE NEXT PAGE ▶▶▶

Model Test 6

3. If $2^{a-2b+4}=3^{b-3}$ then $a - b = ?$

(A) -1   (B) 1   (C) 2   (D) 3   (E) a − b cannot be calculated.

4. $(x^2 - 2x + 1)(x^2 + 2x + 1) = ?$

(A) $(-x^2 - 1)^2$ (B) $(1 - x^2)^2$ (C) $(x^2 - 1)(x^2 + 1)$ (D) $(x^2 + 1)^2 - 2x^2$

(E) None of the above

5. What is the y-intercept of the parabola given by $6x^2 + 4y + 24 = 0$?

(A) (2, 0)   (B) (0, 2)   (C) (0, 6)   (D) (0, − 6)   (E) (− 6, 0)

6. If $\dfrac{1}{3} = y^3$, then $\dfrac{1}{y^9} = ?$

(A) 3   (B) 9   (C) 27   (D) $\dfrac{1}{9}$ (E) $\dfrac{1}{27}$

7. If $(x - y + 2)(x^2 + x + 1) = 0$ then $(2^{y-x})(x^2 - 2xy + 1 + y^2) = ?$

(A) 0.25   (B) 2   (C) 4   (D) 5   (E) 20

8. If $2^{3^2} = 2^{y-1}$, then $y = ?$

(A) 10   (B) 9   (C) 8   (D) 7   (E) 6

GO ON TO THE NEXT PAGE ▶▶▶

9. If $|2 - x| = x - 2$, then which of the following could be the value of x?

(A) – 0.74 (B) – 0.46 (C) 1.12 (D) 1.93 (E) 2.71

10. If $y = \dfrac{3x - 10}{5x + 2}$, then x = ?

(A) $x = \dfrac{3y - 10}{5y + 2}$ (B) $x = \dfrac{2y + 10}{3 - 5y}$ (C) $x = \dfrac{2y + 10}{5 - 3y}$ (D) $x = \dfrac{2 + 10y}{-5 + 3y}$ (E) $x = \dfrac{-2y - 10}{5y + 3}$

11. Which of the following gives all values of x for which $|x + 2| + |3x + 6| < 12$?

(A) -1 < x < 5 (B) -5 < x < -1 (C) -5 < x < 1 (D) 1 < x < 5 (E) -5 < x < 5

12. If x – 2 varies directly as y + 2 and x = 6 when y = 4; then when y = 7, x=?

(A) 5 (B) 6 (C) 7 (D) 8 (E) 9

13. If $i(x + 3) - (y - 4) = i(-5 + 6i)$ where $i = \sqrt{-1}$ then x + y = ?

(A) – 8 (B) – 2 (C) 2 (D) 10 (E) 18

14. Which of the following is not a solution of $-2y + 3x > 12$?

(A) (5, -1) (B) (0,-8) (C) (4, 0) (D) (2, -3) (E) (4, -2)

GO ON TO THE NEXT PAGE ▶▶▶

RUSH	RUSH	RUSH	**Math Level 2 Subject Test**	RUSH	RUSH	RUSH

Model Test 6

15. If $f(g(x)) = e^{-2x-1}$ and $f(x) = e^x$ then $g(x) = ?$

(A) $-2x - 1$ (B) $\ln(x)$ (C) $\ln(2x+1)$ (D) $\dfrac{1}{\ln(x)}$ (E) $\dfrac{1}{2x+1}$

16. If the quantity given by $\dfrac{3r + e - 1}{5r + 6}$ represents a constant for all real values of r, then what is the value of e?

(A) 3/5   (B) 1   (C) -6/5   (D) 23/5   (E) It cannot be determined from the information given

17. Given in figure 1 are three vertices of a parallelogram. Where can be the fourth vertex?

(A) (1, 5)   (B) (-2, 3)   (C) (-3, 2)   (D) (2, -3)   (E) (-1, -5)

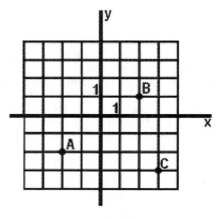

Figure 1

18. Which of the following are correct for the function given partially above?

    I. $g(g(3)) = 1$

    II. $g^{-1}(4) = 3$

    III. $g^2(2) = g(g(2))$

(A) I only   (B) II only   (C) III only   (D) I and II only   (E) I, II and III

x	g(x)
1	2
2	3
3	4
4	1

GO ON TO THE NEXT PAGE ▶▶▶

Model Test 6

19. What is the value of $\dfrac{1+\frac{1}{x}}{\frac{1}{x^2}-1}\cdot\left(1-\frac{1}{x}\right)$ where defined?

(A) 1　　　　　(B) -1　　　　　(C) x　　　　　(D) x − 1　　　　　(E) x + 1

20. A palindromic time is one that reads the same forward as it does backward. For example 12:21 and 4:04 are palindromic times. A 12 hour digital clock displays all times from 12:00 to 11:59. From 12:00 noon to 12:00 noon the next day, how many palindromic times does the digital clock display?

(A) 57　　　　　(B) 90　　　　　(C) 108　　　　　(D) 114　　　　　(E) 180

21. For what values of x is $\dfrac{1}{1+\frac{1}{1+x}}$ undefined?

(A) -2 only　　　(B) -1 only　　　(C) -2 and -1 only　　　(D) -1 and 2 only　　　(E) 1 and 2 only

22. What is the quadratic equation with real coefficients having a root of 3+i where i= $\sqrt{-1}$?

(A) $x^2-6x-10=0$　　(B) $x^2+10x+6=0$　　(C) $x^2+6x+10=0$　　(D) $x^2-10x+6=0$　　(E) $x^2-6x+10=0$

23. $\dfrac{\sin^2 x}{1-\cos x}+\dfrac{\sin^2 x}{1+\cos x}=?$

(A) 2　　　　　(B) -2　　　　　(C) 2cosx　　　　　(D) -2cosx

(E) None of the above

GO ON TO THE NEXT PAGE ▶▶▶

Model Test 6

24. The functions g and h are each periodic with period 5 such that g(3)=4 and h(-2)=8. What is the value of g(h(3)-5)?

(A) -2   (B) 3   (C) 4   (D) 5   (E) 8

25. What is the value of the infinite sum given by $\dfrac{2}{3} - \dfrac{4}{9} + \dfrac{8}{27} - \dfrac{16}{81} + \ldots$?

(A) 2/3   (B) 2/5   (C) 3/2   (D) 5/2   (E) 1

26. The equation given by $y^2 + y = -3x^2 + 2x$ represents

(A) A parabola   (B) An ellipse   (C) A hyperbola   (D) A circle   (E) A pair of straight lines

27. If $3x^2 - px + 2 = 0$ has no real roots, then which of the following must be true of p?

(A) $-2\sqrt{6} \le p \le 2\sqrt{6}$ (B) $-2\sqrt{6} < p < 0$ (C) $-2\sqrt{6} < p < 0$ or $0 < p < 2\sqrt{6}$

(D) $-2\sqrt{6} < p < 2\sqrt{6}$ (E) $0 < p < 2\sqrt{6}$

28. According to the data given in figure 2, x=?

(A) 3.38

(B) 4.94

(C) 4.95

(D) 11.07

(E) 11.08

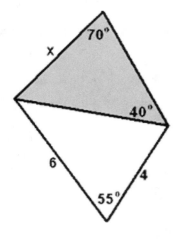

Figure 2

Figure not drawn to scale

GO ON TO THE NEXT PAGE ▶▶▶

29. Graph of f(x − 2) + 1 is given in figure 3 above. Which of the following is the graph of f(x)?

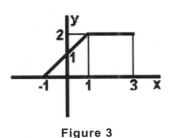

Figure 3

(A)

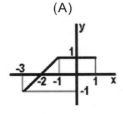

(B)

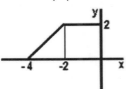

(C)

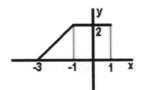

(D)

(E)

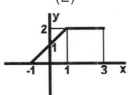

30. If the base radius of a cone is increased by 10%, how must the height of the cone be changed so that the volume of the cone will be decreased by 20%?

(A) Height of the cone must be decreased by 10%.

(B) Height of the cone must be decreased by 66.12%.

(C) Height of the cone must be decreased by 33.88%.

(D) Height of the cone must be decreased by 30%.

(E) Height of the cone need not be changed.

31. At the end of 2003, a certain car was worth $45,000. If the value of the car decreases at a rate of 5 percent each year, approximately how much will the car be worth at the end of 2009?

(A) $36,652   (B) $34,820   (C) $33,079   (D) $32,446   (E) $31,775

32. The quadratic function f(x) is given by $f(x) = ax^2 + bx + c$ where a, b, and c are all real numbers. If f(x) and -f(x) intersect at no points then which of the following must be correct?

(A) $b^2 < 4ac$ (B) $b^2 = 4ac$ (C) $b^2 > 4ac$ (D) $b^2 + 4ac > 0$ (E) $b^2 + 4ac < 0$

GO ON TO THE NEXT PAGE ▶▶▶

Model Test 6

33. It is given that $\mathbf{A}^2 = \begin{bmatrix} -1 & 6 \\ -3 & 2 \end{bmatrix}$ and $\mathbf{A}^3 = \begin{bmatrix} -7 & 10 \\ -5 & -2 \end{bmatrix}$ where **A** is a 2 by 2 square matrix. **A**=?

(A) $\begin{bmatrix} -1 & -2 \\ 1 & -2 \end{bmatrix}$
(B) $\begin{bmatrix} 1 & 2 \\ -1 & 2 \end{bmatrix}$
(C) $\begin{bmatrix} -1 & 2 \\ 1 & -2 \end{bmatrix}$
(D) $\begin{bmatrix} 1 & -2 \\ -1 & 2 \end{bmatrix}$
(E) $\begin{bmatrix} -1 & 2 \\ 1 & 2 \end{bmatrix}$

34. If $f(x) = \dfrac{ax - b}{c}$ and $f(f(x)) = x$ for all x, then which of the following must be true?

   I. $|a| = |c|$

   II. $a \neq 0$

   III. $b > 0$

(A) I only   (B) II only   (C) I and II only   (D) I and III only   (E) I, II and III.

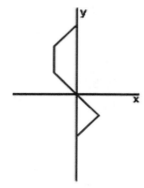

Figure 4

35. Graph of f(x) is rotated 90° clockwise about the origin to give the graph in figure 4 above. What is the graph of the original function f(x)?

(A)   (B)   (C)   (D)   (E)

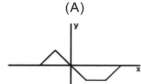 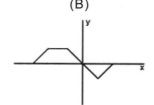 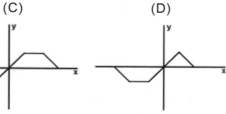 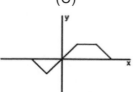 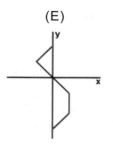

GO ON TO THE NEXT PAGE ▶▶▶

Model Test 6

36. If parallelogram ABCD in figure 5 is rotated about side AB, what is the volume of the solid that is generated?

(A) 243π (B) 270π (C) $300\sqrt{3}\,\pi$

(D) 327π (E) 360π

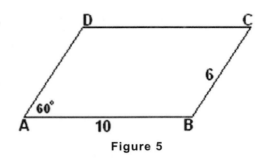

Figure 5

37. If $a = 9^x + 5$ and $b = 3 - 3^x$ then what is a in terms of b?

(A) $3 - b$ (B) $b^2 - 3b$ (C) $b^2 + 4$

(D) $b^2 - 6b + 7$ (E) $b^2 - 6b + 14$

38. In figure 6, the shaded region can be described by which of the following sets of inequalities?

(A) $y < 0$, $x < 3$, $y > -\dfrac{1}{2}x$ (B) $y > 3$, $x < 0$, $y > -2x$

(C) $y < 3$, $x < 0$, $y > -\dfrac{1}{2}x$ (D) $y \leq 3$, $x \leq 0$, $y \geq -\dfrac{1}{2}x$

(E) $y > 3$, $x > 0$, $y \geq -\dfrac{1}{2}x$

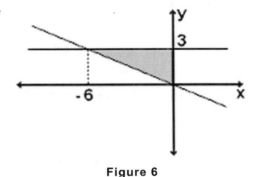

Figure 6
Figure not drawn to scale

39. If $2 \cdot 5^a = 3 \cdot 4^b$ then $\dfrac{a}{b} = ?$

(A) 1.10   (B) 1.11   (C) 1.12   (D) 1.13   (E)1.14

GO ON TO THE NEXT PAGE ▶▶▶

Model Test 6

40. If f is a linear function, f(2) = 5, and f(5) = 2, then f(3)=

(A) − 4   (B) − 3   (C) 3   (D) 4   (E) 7

41. The graph given in figure 7 shows the density d of a solution versus time t PM that a chemist made by mixing exactly two substances A and B with respective densities of d_A=40 gr/cc and d_B=80 gr/cc and d is calculated by the formula $d = \dfrac{d_A \cdot V_A + d_B \cdot V_B}{V_A + V_B}$ where V_A and V_B are the respective volumes of A and B in the solution. Which of the following cannot be inferred from the information given by the graph?

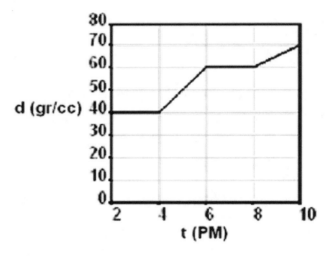

Figure 7

(A) The solution is initially pure.

(B) Between 4:00 PM and 6:00 PM the volume V_B of B added to the mixture is a linear function of time.

(C) Between 6:00 PM and 8:00 PM there are equal amounts of A and B in the mixture.

(D) After 8:00 PM, B is added to the mixture in a slower fashion than that in the 4:00 PM − 6:00 PM period.

(E) If only B is added to the mixture, the density of the solution will never become 80 gr/cc.

| 1. A point | 2. A line | 3. A triangle |
| 4. A quadrilateral | 5. A pentagon | 6. A hexagon |

42. How many of the plane figures given above can be the obtained by the intersection of a plane and a cube?

(A) 2   (B) 3   (C) 4   (D) 5   (E) 6

GO ON TO THE NEXT PAGE ▶▶▶

Model Test 6

43. Which of the following equations represents the line that passes through the points (1,3) and (2,4)?

(A) $\begin{pmatrix} x \\ y \end{pmatrix} = \begin{pmatrix} 1 \\ 3 \end{pmatrix} + t \begin{pmatrix} 2 \\ 4 \end{pmatrix}$ (B) $\begin{pmatrix} x \\ y \end{pmatrix} = \begin{pmatrix} 2 \\ 4 \end{pmatrix} + t \begin{pmatrix} 1 \\ 3 \end{pmatrix}$ (C) $\begin{pmatrix} x \\ y \end{pmatrix} = \begin{pmatrix} 1 \\ 3 \end{pmatrix} + t \begin{pmatrix} 1 \\ 1 \end{pmatrix}$ (D) $\begin{pmatrix} x \\ y \end{pmatrix} = \begin{pmatrix} 4 \\ 2 \end{pmatrix} + t \begin{pmatrix} 1 \\ 1 \end{pmatrix}$ (E) $\begin{pmatrix} x \\ y \end{pmatrix} = \begin{pmatrix} 1 \\ 1 \end{pmatrix} + t \begin{pmatrix} 1 \\ 3 \end{pmatrix}$

44. f(x) = The remainder when [x] is divided by 3, where [x] represents the greatest integer less than or equal to x. If the domain of f(x) is the set of nonnegative real numbers then what is the range of f(x)?

(A) the set of integers

(B) the set of integers less than 3

(C) the set of positive integers less than 3

(D) the set of nonnegative integers less than 3

(E) the set of nonnegative integers less than or equal to 3

45. If f(x) is given by $f(x)=x^3+x+1$, then $(f^{-1} \circ f)(x^2)=?$

(A) x (B) x^2 (C) 1/x (D) $1/x^2$

(E) f(x) is not invertible thus it is impossible to find $(f^{-1} \circ f)(x^2)$.

46. Two sets of data are given as A = {0.1, 2.1, 4.1} and B = {-3.3, -1.3, 0.7}. Which of the following statements are correct?

  I. The mean of the data in set A is greater than that of set B.

  II. The median of the data in set A is greater than that of set B.

  III. The standard deviation of the data in set A is greater than that of set B.

(A) I only (B) II only (C) III only (D) I and II only (E) I, II, and III

GO ON TO THE NEXT PAGE ▶▶▶

Model Test 6

47. Which of the following is the Cartesian representation of the polar equation given by r = cosθ + sinθ?

(A) $x^2 - y^2 = x - y$ (B) $x + y = x^2 + y^2$ (C) $x^2 \cdot y^2 = x \cdot y$ (D) $x^2 \cdot y^2 = x + y$ (E) $- x - y = x^2 + y^2$

48. How many pairs of parallel edges are there in the right regular hexagonal prism given in figure 8?

(A) 15   (B) 18   (C) 24   (D) 33   (E) 36

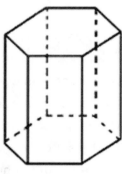

Figure 8

49. Each side of the regular pentagon given in figure 9 is 7 inches long. What is the perimeter of ΔADC in square inches?

(A) 29.6   (B) 22.6   (C) 29.7   (D) 22.7

(E) It cannot be determined from the information given.

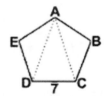

Figure 9

50. Which of the following functions has exactly one horizontal and two vertical asymptotes?

(A) $f(x) = \dfrac{1}{1-x}$ (B) $f(x) = \dfrac{1}{x+1}$ (C) $f(x) = \dfrac{1}{x^2-1}$ (D) $f(x) = \dfrac{x-1}{1-x^2}$ (E) $f(x) = \dfrac{x+1}{x^2-1}$

S T O P

END OF TEST

(Answers on page 199 – Solutions on page 218)

Model Test 7

Test Duration: 60 Minutes

Directions: For each of the following problems, decide which is the **best** of the choices given. If the exact numerical value is not one of the choices, select the choice that best approximates this value. Then fill in the corresponding oval on the answer sheet.

Notes:

- A calculator will be necessary for answering some (but not all) of the questions in this test. For each question you will have to decide whether or not you should use a calculator. The calculator you use must be at least a scientific calculator; programmable calculators and calculators that can display graphs are permitted.

- For some questions in this test you may have to decide whether your calculator should be in the radian mode or the degree mode.

- Figures that accompany problems in this test are intended to provide information useful in solving the problems. They are drawn as accurately as possible **except** when it is stated in a specific problem that its figure is not drawn to scale.

- All figures lie in a plane unless otherwise indicated.

- Unless otherwise specified, the domain of any function **f** is assumed to be the set of all real numbers **x** for which **f(x)** is a real number.

Reference Information: The following information is for your reference in answering some of the questions in this test.

- Volume of a right circular cone with radius **r** and height **h**: $V = \frac{1}{3}\pi r^2 h$

- Lateral area of a right circular cone with circumference of the base **c** and slant height **l**: $S = \frac{1}{2}cl$

- Volume of a sphere with radius **r**: $V = \frac{4}{3}\pi r^3$

- Surface area of sphere with radius **r**: $S = 4\pi r^2$

- Volume of a pyramid with base area **B** and height **h**: $V = \frac{1}{3}Bh$

1. If $1 + \frac{1}{x} = 2 + \frac{2}{x}$ then x=?

(A) – 2 (B) -1 (C) $-\frac{1}{2}$ (D) 0 (E) 1

2. $\left(\frac{1}{b} + \frac{1}{c}\right) : \frac{1}{a}$

(A) $\frac{b+c}{a}$ (B) $\frac{a}{b+c}$ (C) $\frac{1}{abc}$ (D) $\frac{1}{ab} + \frac{1}{ac}$ (E) $\frac{ab+ac}{bc}$

GO ON TO THE NEXT PAGE ▶▶▶

Model Test 7

3. Figure 1 shows the graph of one period of the function y=2Cos(3x). The coordinates of P are

(A) (2, 2π/3)   (B) (2, 3π/2)

(C) (2π/3, 2)   (D) (3π/2, 2)

(E) None of the above

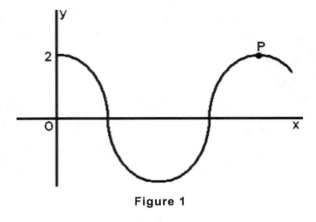

Figure 1

4. Town X is 4 miles to the north of town Y which is 7 miles to the south of town Z. What is the distance between towns X and Z?

(A) 3   (B) 4   (C) 7   (D) 9   (E) 11

5. If $2x + \dfrac{3}{6-3x} = 4 - \dfrac{5}{5x-10}$ then x can be

(A) 2   (B) 3   (C) 4   (D) 8

(E) None of the above

6. If $f(x) = \sqrt{x^2(x+1)}$ and $g(x)=x^2(x+1)$, then g(f(3)) =

(A) 6   (B) 36   (C) 42   (D) 216   (E) 252

GO ON TO THE NEXT PAGE ▶▶▶

Model Test 7

7. In figure 2, r = ?

(A) x·cosθ   (B) y·sinθ   (C) x/sinθ   (D) y/sinθ   (E) x·y

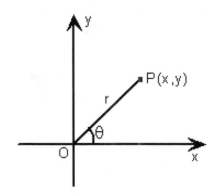

Figure 2

8. If a, b and c are non zero and real numbers and if $5ab^2c^3 = \dfrac{a^2b^2}{c^{-3}}$, then a=

(A) $\sqrt{5}$ (B) 1/5   (C) 5   (D) -5   (E) 25

9. In the triangle in figure 3, if AB=BO, what is the slope of segment BO?

(A) $-\sqrt{3}$ (B) -1   (C) $-\dfrac{\sqrt{3}}{2}$ (D) $\dfrac{\sqrt{3}}{2}$

(E) It cannot be determined from the information given.

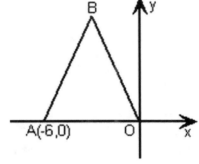

Figure 3

10. Where defined $(\sin x + \cos x)^2 + (\sin x - \cos x)^2 =$?

(A) -2   (B) 0   (C) 2   (D) $2.\sin^2 x$ (E) $2.\cos^2 x$

11. If 5 and -3 are both zeros of the polynomial P(x), then a factor of P(x) is

(A) x + 5   (B) x − 3   (C) $x^2 - 15$ (D) $x^2 - 2x - 15$ (E) $x^2 + 2x - 15$

GO ON TO THE NEXT PAGE ▶▶▶

Model Test 7

12. Eda has a dollars and b cents. She can spend at most how many cents so that she will be left with b dollars and a cents if a and b are distinct nonzero digits?

(A) 0   (B) 792   (C) 891   (D) 990   (E) 999

13. If f(x) = 2x-5 and f(g(10))= 5, which of the following could be g(x)?

(A) x + 5   (B) x − 5   (C) 2x + 5   (D) $\dfrac{x+5}{2}$ (E) $\dfrac{x-5}{2}$

14. If $\angle PBA=45°$, $\angle PAE=60°$, and AE=1; then BA=?

(A) $\sqrt{3}+1$ (B) $\sqrt{3}-1$ (C) $\sqrt{2}-1$ (D) $\sqrt{3}$ (E) $\sqrt{2}$

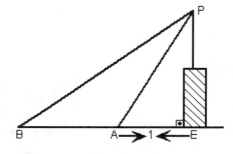

Figure 4
Figure not drawn to scale

15. Figure 6 shows a cube with edge of length 4 centimeters. If points A and B are midpoints of the edges of the cube, what is the area of the shaded region ABCD?

(A) 8.47   (B) 8.94   (C) 16.94   (D) 17.89   (E) 18.71

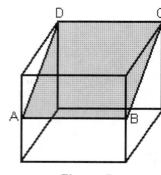

Figure 5

GO ON TO THE NEXT PAGE ▶▶▶

16. If 0<x<π/2 and sinx=0.613, what is the value of cos(x/2)?

(A) 0.249   (B) 0.790   (C) 0.922   (D) 0.946   (E) 0.980

17. An equation of the line m in figure 7 is

(A) x = 0

(B) y = 0

(C) x = -3

(D) y = -3

(E) y = x+3

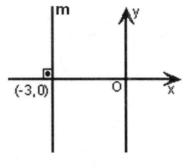

Figure 6

18. Yemlihan's e-mail password is a six digit number that consists of three two digit numbers written one after another. Each of these three numbers satisfies exactly one of the conditions given as follows:

    i. One number is even.

    ii. One number is a multiple of 3.

    iii. One number is the month of the year when Yemlihan was born.

Yemlihan's password can be:

(A) 262710   (B) 112714   (C) 262410   (D) 332613   (E) 332412

19. What is the surface area of a cube inscribed in a sphere with a radius of 4 inches?

(A) 32 square inches   (B) 64 square inches   (C) 128π square inches

(D) 64π square inches   (E) 128 square inches

20. If $f(x)=2x^2 - 3x$, and $g(f(x)) = 4x^2 - 6x+5$, then g(x) can be

(A) 2x   (B) -2x   (C) 2x+5   (D) -2x+5   (E) 2x-5

GO ON TO THE NEXT PAGE ▶▶▶

Model Test 7

21. Which number should be subtracted from each of the three numbers 5, 15 and 50 so that the resulting three numbers form a geometric progression?

(A) – 1   (B) 0   (C) 1   (D) 2   (E) 3

22. If $f(x)=ax^2+bx+c$ for all real number x and if $f(0)= 4$ and $f(1)=3$, then $a+b=$

(A) -1   (B) 0   (C) 1   (D) 2   (E) 3

23. What is the degree measure of the smallest angle of a triangle that has sides of length 40, 48 and 56?

(A) $40.45°$ (B) $42.44°$ (C) $44.42°$ (D) $45.58°$ (E) $54.58°$

24. At how many points do the graphs of the relations $x^2 + y^2 = 16$ and $y = x^2 - 4$ intersect?

(A) 0   (B) 1   (C) 2   (D) 3   (E) 4

25. What is the domain of $f(x) = \dfrac{1}{\sqrt[4]{16 - x^2}}$

(A) All real numbers between -4 and 4 inclusive.   (B) All real numbers.

(C) All real numbers except for -4 and 4.   (D) All real numbers except for 4.

(E) All real numbers between -4 and 4.

GO ON TO THE NEXT PAGE ▶▶▶

Model Test 7

26. If $\sin\theta = \dfrac{1}{\tan\theta}$, which of the following is a possible radian value of θ?

(A) 0   (B) 0.904   (C) 0.905   (D) 5.16   (E) 5.17

27. Figure 8 shows a portion of the graph of $y=\log_2 x$. What is the sum of the areas of the three inscribed rectangles shown?

(A) 7.56

(B) 7.81

(C) 11.17

(D) 12

(E) 12.81

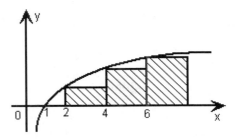

Figure 7

Figure not drawn to scale

28. Which of the following lines are asymptotes of the graph of $y = \dfrac{2-x}{x+1}$

    I: y = -1

    II. x = -1

    III. x = 2

(A) I only   (B) I and II only   (C) I and III only   (D) II and III only   (E) I, II and III only

29. If $f(x-1)= x^2-1$ for all real number x, then f(x)=

(A) x^2 (B) x^2+1 (C) x^2+2x (D) x^2+2x+1 (E) x^2+2x-2

GO ON TO THE NEXT PAGE ▶▶▶

Model Test 7

30. Which of the following could not be the coordinates of a point of a circle tangent to the x-axis and y-axis?

(A) (-3, 0)   (B) (-2, 2)   (C) (0, 0)   (D) (1, 1)   (E) (0, 3)

31. What is the range of the function defined by $f(x) = \begin{cases} -x^2 & x \geq 1 \\ \dfrac{x-1}{x+1} & x < 1 \end{cases}$?

(A) R - {0}   (B) R - {1}   (C) R - {0, 1}   (D) R - [0, 1]
(E) All real numbers

32. If $2x - 3y + 5 = 0$ and $y = 2x^2$ for $x \leq 0$, then x=

(A) -1.09   (B) -0.76   (C) -0.57   (D) -0.07   (E) 1.09

33. If $f(x) = \log_3 x$ for x>0, then $f^{-1}(x)=$

(A) $\dfrac{x}{3}$ (B) $\dfrac{3}{x}$ (C) x^3 (D) 3^x (E) $\log_3 x$

34. If $x_1 = 1$, $x_{n+1} = \sqrt{2 \cdot x_n + 3}$ and $x_k < 2.96$, then which of the following is the greatest value of k?

(A) 2   (B) 3   (C) 4   (D) 5   (E) 6

GO ON TO THE NEXT PAGE ▶▶▶

Model Test 7

35. Figure 9 shows a rectangle inscribed in a semicircle. What is the area of the rectangle in terms of θ and r

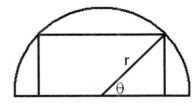

Figure 8

(A) $r \cdot \sin(2\theta)$ (B) $r \cdot \sin\theta \cdot \cos\theta$ (C) $r^2 \cdot \sin\theta$

(D) $2r^2 \cdot \sin\theta$ (E) $r^2 \cdot \sin(2\theta)$

36. In a certain experiment, there is a 0.3 probability that any weighing machine used is in error by more than 10 grams. If 3 weighing machines are used, what is the probability that at least one of them is not in error by more than 10 grams?

(A) 0.027   (B) 0.1   (C) 0.73   (D) 0.973

(E) None of the above

37. If the magnitudes of vectors **a** and **b** are 15 and 8, respectively, then the magnitude of vector **a** − **b** could not be

(A) 6   (B) 8   (C) 10   (D) 12   (E) 14

38. If $4.14^a = 2.07^b$, what is the value of a/b?

(A) 0.21   (B) 0.5   (C) 0.51   (D) 1.95   (E) 2

39. n distinct planes that intersect in a line separate the space into 2n disjoint regions. What is the least number of disjoint regions that can be obtained by introducing a new plane that does not contain the line of intersection?

(A) 2n+1   (B) 2 (n+1)   (C) 3n   (D) 3 (n+1)   (E) 4n

GO ON TO THE NEXT PAGE ▶▶▶

Model Test 7

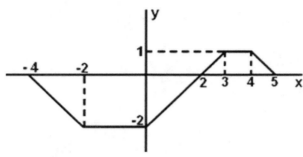

Figure 9

40. Given in figure 10 above is the graph of y = f(x). Which of the following gives the domain and range

of $\dfrac{1}{f(x)}$ correctly?

(A) domain: -4 ≤ x ≤ 5; range: -2 ≤ y ≤ 1

(B) domain: -4 ≤ x ≤ 5; range: -0.5 ≤ y ≤ 1

(C) domain: -4 ≤ x ≤ 5; range: y ≤ -0.5 or y ≥ 1

(D) domain: -4 < x < 2 or 2 < x < 5; range: y ≤ -0.5 or y ≥ 1

(E) domain: -4 < x < 2 or 2 < x < 5; range: y < -0.5 or y > 1

41. If are sin(2x)=$\dfrac{1}{2}$ and $0 \le x \le \dfrac{\pi}{2}$, then x could equal

(A) π/12   (B) π/6   (C) π/4   (D) π/3   (E) π/2

42. sin(θ) + sin(90°-θ) + sin(-θ) + sin(90°+θ)=?

(A) 0   (B) 2sinθ   (C) 2cosθ   (D) sin2θ   (E) 2sinθ + 2cosθ

43. $\dfrac{(n+r+1)!}{(n+r-1)!} = ?$

(A) n+r   (B) n+r+1   (C) (n+r)²   (D) (n+r)²+n+r   (E) (n+r+1)

GO ON TO THE NEXT PAGE ▶▶▶

Model Test 7

44. The radius of the base of a right circular cone is 5 and the perimeter of a cross section perpendicular to the base and passing through the vertex is 36. What is the height of the cone?

(A) 3.6   (B) 6   (C) 7.2   (D) 8   (E) 12

45. What is the converse of the statement given by "If x^2 is divisible by 2, then x+1 is odd"?

(A) x^2 is divisible by 2 or x+1 is odd.

(B) x^2 is divisible by 2 and x+1 is odd.

(C) If x+1 is odd then x^2 is divisible by 2.

(D) If x+1 is not odd then x^2 is not divisible by 2.

(E) If x^2 is not divisible by 2 then x+1 is not odd.

46. Suppose the graph of $f(x) = -x^2 + 3$ is translated 2 units up and 2 units left. If the resulting graph represents g(x), what is the value of g(-2.1)?

(A) -2.1   (B) -4.99   (C) -4.9   (D) 4.9   (E) 4.99

47. In how many ways can 9 identical apples be partitioned into two identical baskets?

(A) 5   (B) 9   (C) 10   (D) 36   (E) 72

48. Which of the following has an element that is greater than any other element in that set?

     I. The set of negative real numbers.

     II. The set of prime numbers less than 100.

     III. The set of positive numbers x such that $| x - 1 | \le 100$

(A) None   (B) I only   (C) I and III only   (D) II and III only   (E) I, II and III

GO ON TO THE NEXT PAGE ▶▶▶

Model Test 7

49. What is the length of the minor axis of the ellipse whose equation is $\dfrac{x^2}{50} + \dfrac{y^2}{20} = \dfrac{1}{10}$?

(A) 1.41   (B) 2.24   (C) 2.83   (D) 3.16   (E) 7.07

50. Under which of the following conditions is $\dfrac{a+b}{ab}$ always negative?

(A) $0 < a < b$ (B) $b < 0 < a$ (C) $a < b < 0$ (D) $a < -1$ and $0 < b < 1$

(E) None of the above.

S T O P

END OF TEST

(Answers on page 199 – Solutions on page 220)

Model Test 8

Test Duration: 60 Minutes

Directions: For each of the following problems, decide which is the **best** of the choices given. If the exact numerical value is not one of the choices, select the choice that best approximates this value. Then fill in the corresponding oval on the answer sheet.

Notes:

- A calculator will be necessary for answering some (but not all) of the questions in this test. For each question you will have to decide whether or not you should use a calculator. The calculator you use must be at least a scientific calculator; programmable calculators and calculators that can display graphs are permitted.

- For some questions in this test you may have to decide whether your calculator should be in the radian mode or the degree mode.

- Figures that accompany problems in this test are intended to provide information useful in solving the problems. They are drawn as accurately as possible **except** when it is stated in a specific problem that its figure is not drawn to scale.

- All figures lie in a plane unless otherwise indicated.

- Unless otherwise specified, the domain of any function **f** is assumed to be the set of all real numbers **x** for which **f(x)** is a real number.

Reference Information: The following information is for your reference in answering some of the questions in this test.

- Volume of a right circular cone with radius **r** and height **h**: $V = \frac{1}{3}\pi r^2 h$

- Lateral area of a right circular cone with circumference of the base **c** and slant height **l**: $S = \frac{1}{2}cl$

- Volume of a sphere with radius **r**: $V = \frac{4}{3}\pi r^3$

- Surface area of sphere with radius **r**: $S = 4\pi r^2$

- Volume of a pyramid with base area **B** and height **h**: $V = \frac{1}{3}Bh$

1. If $\frac{3}{2x+1} = 0$ then x =?

(A) -0.5　　　(B) -1　　　(C) -2　　　(D) 0　　　(E) There is no such value of x.

2. If $x = \log_2 9$ then what is the value of x rounded to the nearest hundredth?

(A) 3.14　　　(B) 3.15　　　(C) 3.16　　　(D) 3.17　　　(E) 3.18

GO ON TO THE NEXT PAGE ▶▶▶

Model Test 8

3. The number $(-243)^{\frac{-1}{5}}$ equals

(A) -3   (B) $\frac{-1}{3}$ (C) $\frac{1}{3}$ (D) 3   (E) an undefined quantity

4. $\dfrac{(x^{-2})^{-5}}{x^2 \cdot x^5} = ?$

(A) $\dfrac{1}{x^3}$ (B) $\dfrac{1}{x}$ (C) 1   (D) x   (E) x^3

5. What is the solution set of the inequality given by $|3x-1| \geq 7$?

(A) x < -2 or x > $\frac{8}{3}$ (B) x ≤ -2 or x ≥ $\frac{8}{3}$ (C) -2 ≤ x ≤ $\frac{8}{3}$ (D) x ≤ -8 or x ≥ $\frac{2}{3}$ (E) -8 ≤ x ≤ $\frac{2}{3}$

6. If the square root of the cube of p is 10, then what is the value of p rounded to the nearest tenth?

(A) 4.5   (B) 4.6   (C) 4.7   (D) 32.6   (E) 1000000.0

7. The algebraic expression given by $\dfrac{\dfrac{2x-1}{2x} - 1}{2x - \dfrac{4x^2-1}{2x}}$ can be simplified to

(A) 1   (B) − 1   (C) 2x   (D) − 2x   (E) − 2x − 1

GO ON TO THE NEXT PAGE ▶▶▶

Model Test 8

8. What are all intervals on which $\dfrac{x+1}{x^2-9}$ is positive?

(A) -3 < x ≤ -1 or x > 3　　　　(B) -3 < x < -1 or x > 3　　　　(C) x < -3 or -1 ≤ x < 3

(D) x < -3 or -1 < x < 3　　　　(E) x < -1 or x > 3

9. The equation given by $y^2 + 4y = x^2 + 2x$ represents

(A) A circle　　　(B) An ellipse　　　(C) A parabola　　　(D) A hyperbola　　　(E) A pair of straight lines

10. For the right square pyramid given in figure 1, what is the measure of angle DEB rounded to the nearest degree in the shaded triangle?

(A) 23　　　(B) 44　　　(C) 46　　　(D) 47　　　(E) 67

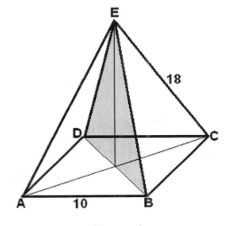

Figure 1

Figure not drawn to scale

11. Graph of f(x) is given in figure 2. What will be the graph of f(|x|)?

(A)

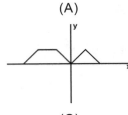

(B)

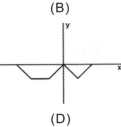

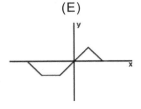

Figure 2

(C)

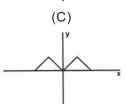

(D)

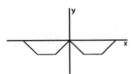

(E)

GO ON TO THE NEXT PAGE ►►►

12. If g(x, y) = B satisfies the relations given by g(x, y) = g(-x, y) = g(-x, -y) = B then the curve described by g(x, y) = B is

    I. symmetric with respect to the x axis

    II. symmetric with respect to the y axis

    III. symmetric with respect to the origin

(A) I only   (B) II only   (C) III only   (D) II and III only   (E) I, II and III

13. What is the period of the function given by $f(x) = 3 \cdot \sin^2(2x) + 1$?

(A) $\dfrac{\pi}{2}$ (B) π (C) 2π (D) 3π (E) 4π

14. Which of the following is the equation of a circle?

(A) $(x - 1)^2 + (y + 1)^2 + 1 = 0$ (B) $x^2 + 4x + y^2 + 2y = 0$ (E) $y = \sqrt{9 - (x + 4)^2}$

(C) $x^2 + 4x - y^2 + 2y = 0$ (D) $x^2 - 4x + 4 + y^2 = 0$

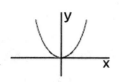

Figure 3

15. Graph of f(x) is given in figure 3 above. The graph of the slopes of tangents to f(x) will most likely resemble which of the following?

(A)   (B)   (C)   (D)   (E)

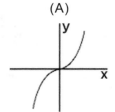 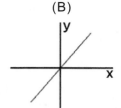 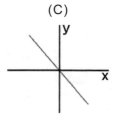 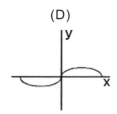 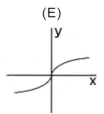

GO ON TO THE NEXT PAGE ▶▶▶

Model Test 8

16. For which of the following values of x is $f(x) = e^{-x} \cdot (2x^3 - 4x + 1)$ less than 1?

(A) -1.43   (B) 2.77   (C) 3.26   (D) 4.15   (E) 7.32

17. Two pools identical in volume are filled to capacity and each pool will be drained by a pipe that can empty it completely in 5 hours. If the pools are started to be unfilled on the same day at 11.00 AM and 1.00 PM respectively, at what time will the second pool contain exactly five times as much water as the first one?

(A) 1.30 PM   (B) 2.00 PM   (C) 2.30 PM   (D) 3.00 PM   (E) 3.30 PM

18. What is the domain of the function given by $f(x) = \dfrac{4}{|2x - 6| - 2}$?

(A) All real numbers except 2 and 4   (B) $x > 0$ or $x \leq -2$

(C) All real numbers except -2 and -4 (D) All real numbers

(E) All real numbers except 3

Figure 4

19. The graph given in figure 4 above is the solution set of which of the following inequalities?

(A) $|x - 9| < 6$ (B) $|x - 6| > 9$ (C) $|x - 6| < 9$ (D) $|x - 9| \geq 6$ (E) $|x + 6| \geq 9$

20. If (x, y) are the points on the curve given by $y = x^3 - 2x + 1$, what will be the equation of the reflection of this curve about the origin?

(A) $y = -x^3 + 2x - 1$ (B) $y = -x^3 + 2x + 1$ (C) $y = x^3 - 2x - 1$

(D) $x = y^3 - 2y + 1$ (E) $x = y^3 - 2y - 1$

GO ON TO THE NEXT PAGE ▶▶▶

Model Test 8

21. What is the solution set of the equation given by $(x^2 - x)(-2x - 1) - (x^2 + x)(2x + 1) = 0$?

(A) $\left\{\dfrac{-1}{2}, \dfrac{1}{2}, 0\right\}$ (B) $\left\{\dfrac{-1}{2}, 0, 1\right\}$ (C) $\left\{\dfrac{-1}{2}, 0\right\}$ (D) $\left\{\dfrac{-1}{2}, \dfrac{1}{2}\right\}$ (E) $\left\{\dfrac{1}{2}, 0\right\}$

22. As is given in figure 5, a man whose eye level is at position E sights a bird at position B at the angle of elevation of $\theta = 14°$. At this instant sunlight is oblique making an angle of $\beta = 22°$ with the vertical. If the man's eyes are 5.5 feet above the ground and the bird is flying at an altitude of BV = 20 feet, what is the distance between the bird's shadow S and the man's feet F?

(A) 86 feet and 2.84 inches

(B) 66.24 inches

(C) 66 feet and 2.84 inches

(D) 66 feet and 0.24 inches

(E) 86 feet and 3 inches

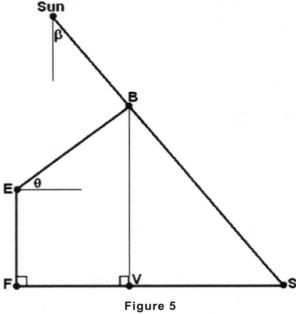

Figure 5
Figure not drawn to scale

23. $(\sec x - 1)(\sec x + 1) = ?$

(A) $\tan^2 x$ (B) $\cot^2 x$ (C) $\sin^2 x$ (D) $\cos^2 x$ (E) $-\cos^2 x$

24. In a geometric sequence, the common ratio is -2 and the 6th tem is -144. What is the second term?

(A) 18   (B) -9   (C) 9   (D) -18   (E) -4.5

GO ON TO THE NEXT PAGE ▶▶▶

Model Test 8

25. What are all values of x for which $\tan(x) = \dfrac{c}{3}$ and $\sin(x) = \dfrac{c}{\sqrt{9+c^2}}$ in the interval $0 < x < 2\pi$?

(A) $0 < x < \dfrac{\pi}{2}$ or $\dfrac{3\pi}{2} < x < 2\pi$ (B) $0 < x < 2\pi$ (C) $0 < x < \dfrac{\pi}{2}$

(D) $0 < x < \dfrac{\pi}{2}$ or $\pi < x < \dfrac{3\pi}{2}$ (E) $0 < x < \pi$

26. For the polynomial P(x) it is given that $P(x - 1) = x^3 - 2x^2 + 3hx$ and when P(x) is divided by (x + 2) the remainder is 12. h = ?

(A) -16   (B) 8   (C) 5   (D) - 5   (E) -8

27. What is the maximum number of lines that can be obtained when 10 distinct planes intersect in such a way that no line belongs to more than two planes?

(A) 100   (B) 90   (C) 55   (D) 45   (E) 36

28. If $f(x) = \sqrt{x-3}$ then domain of f(x) cannot include which of the following numbers?

(A) 3   (B) π (C) $\sqrt{5}$ (D) 2e   (E) 6

29. How many distinct sums can be obtained by adding any two different numbers selected from the elements of set A or B where A = {1, 2, 4, 8, 16, 32, 64, 128} and
B = {1, 2, 4, 8}?

(A) 4·8   (B) $2^8 - 1$ (C) $\dbinom{8}{2}$ (D) $\dbinom{4}{2} \cdot \dbinom{8}{2}$ (E) $\dbinom{4}{2} + \dbinom{8}{2}$

GO ON TO THE NEXT PAGE ▶▶▶

Model Test 8

30. Three student buses A, B and C, depart from *RUSH* Academy heading non stop with constant speeds toward the math contest in Mathville. When A reaches Mathville, B and C have 80 and 144 miles to go and when B reaches Mathville, C still has 80 miles to go. What is the distance in miles between *RUSH* Academy and Mathville?

(A) 300   (B) 350   (C) 400   (D) 450

(E) There is not enough information to solve this problem.

$$1, 2, 6, 15, 31,\dots$$

31. The terms in the sequence above are generated by using a cubic function whose domain is consecutive integers. What is the seventh term of this sequence?

(A) 56   (B) 67   (C) 74   (D) 92   (E) 98

32. Which of the following can be the equation of the parabola given in figure 6?

(A) $x = y^2 + 12$

(B) $x = 3(y - 2)^2$

(C) $x = 2(y - 3)^2$

(D) $y = 5\sqrt{x+1} - 3$

(E) $y = (y - 2)^2 + 12$

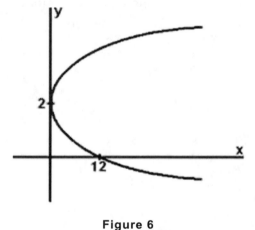

Figure 6
Figure not drawn to scale

33. What is the locus of points at a distance of 5 inches from a plane and at a distance of 7 inches from a point on this plane?

(A) Two planes   (B) A plane   (C) A sphere   (D) A circle   (E) Two circles

GO ON TO THE NEXT PAGE ▶▶▶

Model Test 8

34. Given in figure 7 is a quadratic function of the form $f(x) = x^2 + mx + n$.

What are the coordinates of the vertex of this parabola?

(A) (1.5, -0.25)   (B) (-0.25, 1.5)   (C) (2, -0.25)

(D) (2, 0)   (E) (0, 2)

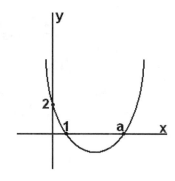

Figure 7

Figure not drawn to scale

35. Given in figure 8 are two periods of a periodic function $f(x)$. What is the value of $f(140) + f(142)$?

(A) 1   (B) 2   (C) 3   (D) 12

(E) None of the above

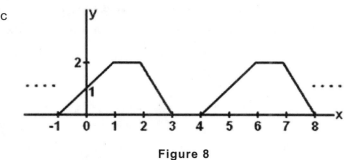

Figure 8

36. Which of the following equations is equivalent to $y = 2x$?

   I. $y^2 = 4x^2$ II. $y^3 = 8x^3$ III. $\sqrt{y} = \sqrt{2x}$

(A) I only   (B) II only   (C) III only   (D) II and III only   (E) I, II and III

37. If $g(x) = f(x - 2) + 1$ then how are the graphs of $f(x)$ and $g(x)$ related?

(A) Graph of $f(x)$ can be obtained by shifting graph of $g(x)$ by 2 units left and 1 unit up.

(B) Graph of $f(x)$ can be obtained by shifting graph of $g(x)$ by 2 units right and 1 unit up.

(C) Graph of $f(x)$ can be obtained by shifting graph of $g(x)$ by 1 unit right and 2 units up.

(D) Graph of $f(x)$ can be obtained by shifting graph of $g(x)$ by 1 unit left and 2 units down.

(E) Graph of $f(x)$ can be obtained by shifting graph of $g(x)$ by 2 units left and 1 unit down.

GO ON TO THE NEXT PAGE ▶ ▶ ▶

38. What are all asymptotes of the function given by $f(x) = \dfrac{x^2}{x^2-1}$?

(A) Vertical asymptote at x = 1; no horizontal asymptote.

(B) Vertical asymptote at x = 1; horizontal asymptote at y = 1.

(C) Vertical asymptotes at x = 1 and x = − 1; no horizontal asymptote.

(D) Horizontal asymptotes at x = 1 and x = − 1; vertical asymptote at y = 1.

(E) Vertical asymptotes at x = 1 and x = − 1; horizontal asymptote at y = 1.

39. A square is inscribed in the circle $(x - 1)^2 + (y + 1)^2 = 8$. If one vertex of this square is the point (3, -3) then which of the following can be another vertex of this square?

(A) (-1, 1)   (B) (-1, 3)   (C) (1, 1)   (D) (-1, -1)   (E) (3, -1)

40. What is the center of symmetry of the hyperbola given by $y = \dfrac{2x+7}{x+4}$?

(A) (3.5, 4)   (B) (2, - 4)   (C) (- 4, 2)   (D) (- 4, -2)   (E) (- 4, -3.5)

41. What is the inverse of the function given by $f(x) = e^{2x} + 1$

(A) $f^{-1}(x) = \dfrac{1}{2} \cdot \ln(x-1)$ (B) $f^{-1}(x) = \dfrac{1}{2} \cdot \ln(x+1)$ (C) $f^{-1}(x) = \dfrac{-1}{2} \cdot \ln(x+1)$

(D) $f^{-1}(x) = \dfrac{-1}{2} \cdot \ln(x-1)$ (E) $f^{-1}(x) = \ln(x-2)$

GO ON TO THE NEXT PAGE ►►►

Model Test 8

42. There are eight pairs of socks in a bag and every pair has a different color. Mete needs a pair of socks where the individual socks have the same color. In order to make sure, at least how many individual socks must he withdraw from the bag?

(A) 8   (B) 9   (C) 10   (D) 14   (E) 16

43. The numbers in the set {1, 2, 3, 4, 5, 6, 7, 8, 9} are written on separate balls and placed in a bag. If two balls are randomly selected from the bag, what is the probability that the sum of the numbers on the balls will be greater than 6?

(A) $\frac{5}{6}$ (B) $\frac{1}{9}$ (C) $\frac{22}{27}$ (D) $\frac{16}{81}$ (E) $\frac{1}{6}$

44. If $f(x) = 2^{2x+1} - 1$ and $g(x) = 4^x + 1$, then what is $g(x)$ in terms of $f(x)$?

(A)   (B)   (C)   (D)   (E)

$g(x) = \frac{f(x)+2}{3}$ $g(x) = \frac{f(x)+3}{2}$ $g(x) = \frac{f(x)-3}{2}$ $g(x) = \frac{-f(x)+3}{2}$ $g(x) = \frac{f(x)+3}{-2}$

45. If $f(2x - 1) = 3x + 4$ and $f(A) = B$ then what is the relation between A and B?

(A) B = 1.5A − 5.5   (B) B = 1.5A + 5.5   (C) A = 1.5B + 5.5   (D) A = 1.5B − 5.5   (E) A = 5.5B + 1.5

46. What is the maximum value of $\sin(2x) \cdot \cos(2x)$?

(A) 0.25   (B) 0.5   (C) 1   (D) 2   (E) 4

GO ON TO THE NEXT PAGE ▶▶▶

Model Test 8

47. If tanA = $\frac{1}{4}$·cot(10°) and A is an acute angle then what is the measure of A?

(A) 0.957   (B) 1.418   (C) 2.5°   (D) 20°   (E) 54.8

48. It is given that sin²x < cosx and 0° ≤ x ≤ 360°. What is the solution set for x?

(A) 0° < x < 51.8° or 308.2° < x < 360°   (B) 51.8° < x < 308.2°

(C) 0° ≤ x < 51.8° or 308.2° < x ≤ 360°   (D) 51.8° ≤ x ≤ 308.2°

(E) 0° ≤ x ≤ 51.8° or 308.2° ≤ x ≤ 360°

49. Point A(a, b) is reflected across the x axis to give point B. Point B is then reflected across the line y = - x to give point C. Finally point C is reflected across the y axis to give point D. What is the distance between the points A and D?

(A) 2a   (B) 2b   (C) |2a| (D) |2b| (E) $\sqrt{2\cdot(a+b)^2}$

50. According to the data given in figure 9 what is the area of the shaded triangle?

(A) 4.178   (B) 13.690   (C) 16.713

(D) 16.712   (E) 17.457

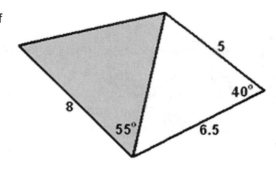

Figure 9

Figure not drawn to scale

S T O P

END OF TEST

(Answers on page 199 – Solutions on page 223)

Model Test 9

Test Duration: 60 Minutes

Directions: For each of the following problems, decide which is the **best** of the choices given. If the exact numerical value is not one of the choices, select the choice that best approximates this value. Then fill in the corresponding oval on the answer sheet.

Notes:

- A calculator will be necessary for answering some (but not all) of the questions in this test. For each question you will have to decide whether or not you should use a calculator. The calculator you use must be at least a scientific calculator; programmable calculators and calculators that can display graphs are permitted.

- For some questions in this test you may have to decide whether your calculator should be in the radian mode or the degree mode.

- Figures that accompany problems in this test are intended to provide information useful in solving the problems. They are drawn as accurately as possible **except** when it is stated in a specific problem that its figure is not drawn to scale.

- All figures lie in a plane unless otherwise indicated.

- Unless otherwise specified, the domain of any function **f** is assumed to be the set of all real numbers **x** for which **f(x)** is a real number.

Reference Information: The following information is for your reference in answering some of the questions in this test.

- Volume of a right circular cone with radius **r** and height **h**: $V = \frac{1}{3}\pi r^2 h$

- Lateral area of a right circular cone with circumference of the base **c** and slant height **l**: $S = \frac{1}{2}cl$

- Volume of a sphere with radius **r**: $V = \frac{4}{3}\pi r^3$

- Surface area of sphere with radius **r**: $S = 4\pi r^2$

- Volume of a pyramid with base area **B** and height **h**: $V = \frac{1}{3}Bh$

1. The set of all ordered pair (x,y) that satisfy the system $\begin{cases} y + x = 0 \\ xy + 1 = 0 \end{cases}$ is

(A) {(1,1)}   (B) {(-1,-1)}   (C) {(1,-1)}   (D) {(1,-1),(1,1)}   (E) {(1,-1), (-1,1)}

2. If -1<k<0, which of the following is greater than 1?

(A) −k   (B) 2k   (C) k^2 (D) $-\frac{1}{k}$ (E) $\frac{1}{k}$

GO ON TO THE NEXT PAGE ▶▶▶

Model Test 9

3. When a certain number A is divided by 4, the remainder is 2. What is the remainder when the square of A is divided by 8?

(A) 0   (B) 2   (C) 4   (D) 6   (E) It cannot be determined from the information given.

4. If $f(x)= -3x^2+kx+4$ and $f(1)=f(-1)$, then k=

(A) -1   (B) 0   (C) 1   (D) 3   (E) 4

5. If functions f, g and h are defined by $f(x)=2x+1$, $g(x)=\dfrac{x-1}{2}$ and $h(x)=3x$, then $f(g(h(5)))=$

(A) 7   (B) 10   (C) 15   (D) 16   (E) 20

6. For a 240 mile trip with two laps of equal length, a car is driven at an average speed of 60 miles per hour during the first lap. If the average speed for the entire trip is 72 miles per hour, what is the average speed of the car for the second lap?

(A) 48   (B) 66   (C) 80   (D) 84   (E) 90

7. If $f(x) = x^2-1$, then $f(1-x)=$

(A) x(x-1)   (B) x(x-2)   (C) (x-1)(x+1)   (D) (x-1)(x+2)   (E) –x(x+2)

8. Which of the following choices best describes the points (x, y) on the Cartesian plane that satisfy the inequalities given by: $x \geq -2$; $y \leq 10$; and $y \geq 3x - 2$?

(A) The points on a triangle   (C) The points inside a triangle

(B) The points outside a triangle   (D) The points inside and on a triangle

(E) The points outside and on a triangle

GO ON TO THE NEXT PAGE ▶▶▶

9. In figure 1, if AB is a diameter of the circle and $\cos x = \dfrac{4}{5}$, what is the value of siny?

(A) 1/5　　　(B) 2/5　　　(C) 3/5　　　(D) 4/5　　　(E) 1

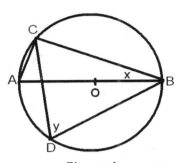

Figure 1
Figure not drawn to scale

10. The shaded region in figure 2 is bounded by the following lines except

(A) x=1　　(B) x=2　　(C) y=0　　(D) 3x+2y=6　　(E) 2x+3y=6

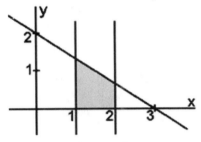

Figure 2

11. If a right triangle is rotated 360° about one of its legs, the solid generated will be a

(A) cube　　　　(B) hemisphere　　　(C) cone　　　　(D) cylinder　　　(E) sphere

x	g(x)
1	2
2	3
3	4
4	1

12. Which of the following are correct for the function given partially above?

　I. g(g(1)) = 3

　II. g(g^{-1}(1)) = 1

　III. g^2(1) = g(g(1))

(A) I only　　　　(B) II only　　　　(C) III only　　　　(D) I and II only　　　(E) I, II and III

GO ON TO THE NEXT PAGE ▶▶▶

Model Test 9

13. An angle measure of $\dfrac{5\pi}{6}$ radians is equivalent to an angle measure of

(A) 105°   (B) 120°   (C) 135°   (D) 150°   (E) 180°

14. Between 0 and 1 on the number line, marks are placed at every fourths and every fifths. Which of the following is not a distance between two adjacent marks?

(A) $\dfrac{1}{20}$ (B) $\dfrac{1}{10}$ (C) $\dfrac{3}{20}$ (D) $\dfrac{1}{5}$ (E) $\dfrac{1}{4}$

15. The solution set of $\dfrac{x^2-1}{x+1} \geq 0$ is

(A) x > 1   (B) x ≥ 1   (C) x ≥ 0   (D) the empty set   (E) x is any real number

16. In figure 3, the shaded region is outside a circle and inside an ellipse. If the ratio of the shaded region to the area of the circle is $\dfrac{3}{4}$, what is the ratio of a to b?

(A) $\dfrac{4}{7}$ (B) π (C) $\dfrac{4}{3}$ (D) $\dfrac{7}{4}$ (E) 2π

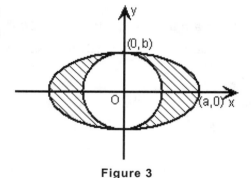

Figure 3

17. Where defined $\dfrac{\sec\theta}{\sin\theta} = ?$

(A) 1   (B) $\csc\theta$ (C) $\tan\theta$ (D) $\sin\theta.\cos\theta$ (E) $\dfrac{1}{\sin\theta.\cos\theta}$

GO ON TO THE NEXT PAGE ▶▶▶

Model Test 9

18. In figure 4, a cut made parallel to the base of the bigger cone slices it into two segments in such a way that $\dfrac{r}{R}=\dfrac{2}{3}$. What is the ratio of the volume of the upper segment to that of the lower segment?

(A) $\dfrac{8}{19}$　　(B) $\dfrac{2}{3}$　　(C) 2　　(D) $\dfrac{4}{9}$　　(E) $\dfrac{4}{5}$

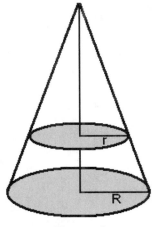

Figure 4

19. $A_{4\times5}$ and $B_{5\times2}$ are both matrices in the matrix multiplication given by **C=AB**. How many rows and columns does **C** have?

(A) 4 rows and 5 columns　　(B) 4 rows and 2 columns　　(C) 5 rows and 2 columns

(D) 5 rows and 4 columns　　(E) 2 rows and 4 columns

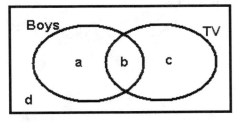

	Boys	Girls	Totals
TV		9	
Other than TV	7		
Totals		15	

20. The data given by the graph and the table above represents the leisure time activities of the senior class in **RUSH** Academy. If there are totally 40 people in the senior class then what is the value of b+d?

(A) 13　　(B) 15　　(C) 16　　(D) 24　　(E) 27

21. If $0 < \theta \le \dfrac{3\pi}{2}$ and $\sin\dfrac{\pi}{6}=\cos\left(\dfrac{\pi}{6}+\theta\right)$, then θ is

(A) $\pi/4$　　(B) $\dfrac{3\pi}{4}$　　(C) $\dfrac{5\pi}{6}$　　(D) $\dfrac{4\pi}{3}$　　(E) $\dfrac{3\pi}{2}$

GO ON TO THE NEXT PAGE ►►►

Model Test 9

22. If $f(x) = \dfrac{x+10}{(x^2+10)(x-10)}$, for what value of is x is f(x) undefined?

(A) -10　　　　(B) -5　　　　(C) 0　　　　(D) 5　　　　(E) 10

23. If, for all x, $2^x + 2^x + 2^x + 2^x = k \cdot 2^{x-1}$, then k=

(A) 1　　　　(B) 3　　　　(C) 4　　　　(D) 8　　　　(E) 16

24. If $f(x) = (x-2)^2 + 3$, what is the minimum value of the function f?

(A) -2　　　　(B) 0　　　　(C) 1　　　　(D) 2　　　　(E) 3

25. If the line x=k-1 is tangent to the circle $x^2 + (y-3)^2 = 16$, then k=

(A) -4 or 4　　　　(B) -3 or 5　　　　(C) -1 or 7　　　　(D) 0 or 3　　　　(E) 0 or 8

26. If $f(x) = 2x^2$, which of the following is the graph of y=f(-x)?

(A)　　　　(B)　　　　(C)　　　　(D)　　　　(E)

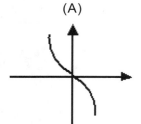 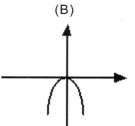 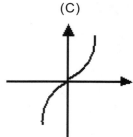 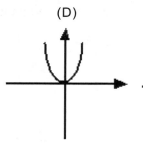 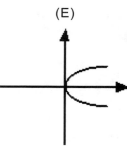

GO ON TO THE NEXT PAGE ▶▶▶

Model Test 9

27. In a competition, the mean score was 80, the median score was 85, and the standard deviation of the scores was 14. The administration added 20 points to each score due to a regulation. Which of the following statements is correct for the new scores?

    I. The new mean score is 100.

    II. The new median score is 105.

    III. The new standard deviation of the scores is 14.

(A) I only   (B) II only   (C) III only   (D) I and II only   (E) I, II and III

28. If x-1 is a factor of $3x^3-2x^2+kx+5$, then k is

(A) − 6   (B) -2   (C) 0   (D) 5   (E) 6

29. If $f(x)=x^2+3x-10$, then the set of all k for which f(-k)=f(k) is

(A) − 5   (B) 0   (C) 2   (D) 5   (E) All real numbers

30. If $\sin\theta > -\cos\theta$ and $0 \leq \theta \leq 2\pi$, then θ could be

(A) $\dfrac{3\pi}{2}$ (B) $\dfrac{7\pi}{6}$ (C) $\dfrac{5\pi}{4}$ (D) π (E) $\dfrac{2\pi}{3}$

31. Which of the following could be the graph of the set of all pairs $\left(\dfrac{1}{x},\dfrac{1}{y}\right)$, where x=secθ, y=cscθ,

and $0 \leq \theta \leq 2\pi$?

(A)   (B)   (C)   (D)   (E)

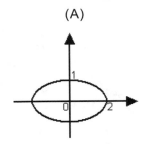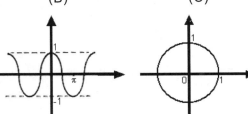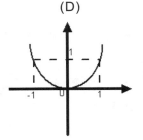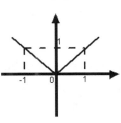

GO ON TO THE NEXT PAGE ▶▶▶

Model Test 9

32. If for all x, $f(x)=a^x$ and $\dfrac{f(x+1)}{f(x)}=8$, then a=

(A) 1 (B) 2 (C) 4 (D) 8 (E) 16

33. Which of the following could be an equation of the graph shown in figure 5?

(A) $y = \sin(x) - 1$ (B) $y = \cos(x) - 1$

(C) $y = \csc(x) + 1$ (D) $y = \csc(x) - 1$

(E) $y = \sec(x) - 1$

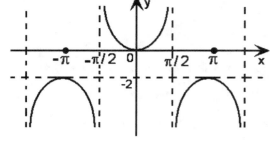

Figure 5

34. If $\cos x=\dfrac{\sqrt{2}}{2}$ and $0 \le x \le \pi$, then $\cos(3x)=$

(A) $-\dfrac{1}{2}$ (B) $-\dfrac{\sqrt{2}}{2}$ (C) $\dfrac{\sqrt{2}}{2}$ (D) $\dfrac{3\sqrt{2}}{2}$ (E) $\dfrac{1}{2}$

35. $\dfrac{n!+(n-1)!}{(n+2)!-(n+1)!}=$

(A) $2n-1$ (B) $\dfrac{1}{n^2+n}$ (C) $\dfrac{1}{n+1}$ (D) $\dfrac{n-1}{n+1}$ (E) $\dfrac{n^2}{n-1}$

36. What is the range of the function defined by $f(x)=1+\dfrac{1}{1-\dfrac{1}{x}}$?

(A) All real numbers except for 0   (B) All real numbers except 0 and 1

(C) All real numbers except for 1   (D) All real numbers except 1 and 2

(E) All real numbers except for 2

GO ON TO THE NEXT PAGE ▶▶▶

Model Test 9

37. If a < b and b < c, which of the following must be true?

   I. −c<a   II. a+c=0   III. a+b < b+c

(A) I only   (B) II only   (C) III only   (D) I and III only   (E) I, II and III

38. If $\sum_{k=0}^{7}(k-2) = \sum_{k=0}^{7}k + A$, then A=

(A) -14   (B) -16   (C) 8   (D) 14   (E) 16

39. Among the eight vertices of a cube how many different sets of three can be selected to construct an equilateral triangle?

(A) 8   (B) 12   (C) 18   (D) 24   (E) More than 24

40. In the function given by f(x) = k·x, x is an arbitrarily chosen positive three digit integer and the arithmetic mean of the digits of f(x) is the same as the arithmetic mean of the digits of x. Which of the following can be the number k?

(A) 10   (B) 11   (C) 101   (D) 110   (E) 1001

41. If two fair dice are tossed, what is the probability that the difference of the number of dots on the top faces will be 4?

(A) $\frac{1}{3}$ (B) $\frac{1}{4}$ (C) $\frac{1}{6}$ (D) $\frac{1}{9}$ (E) $\frac{1}{18}$

GO ON TO THE NEXT PAGE ▶▶▶

Model Test 9

42. What is $\lim\limits_{n\to\infty} \dfrac{n+\sqrt{4n^2+1}}{2n-1}$

(A) -1   (B) 1   (C) 1.5   (D) 5

(E) The limit does not exist.

43. The least positive integer N for which each of $\dfrac{N}{54}$, $\dfrac{N}{55}$, and $\dfrac{N}{56}$ is an integer is

(A) $2^3 \cdot 3^3 \cdot 5 \cdot 7 \cdot 11$ (B) $2^4 \cdot 3^3 \cdot 5 \cdot 7 \cdot 11$ (C) $11 \cdot 10 \cdot 9 \cdot 8 \cdot 7 \cdot 6$ (D) $11 \cdot 9 \cdot 8 \cdot 7 \cdot 6 \cdot 5$

(E) N cannot be determined using the given information.

44. In figure 6, a cube with an edge of length 1 is given. A triangle with an area of $\dfrac{\sqrt{3}}{2}$ has two vertices of A and C. What is the third vertex?

(A) B

(B) D

(C) F

(D) G

(E) H

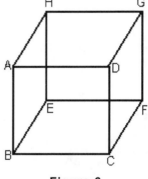

Figure 6

45. i^{4n+k} is a negative real number for all positive integer values of n if k is

(A) 46   (B) 47   (C) 48   (D) 52   (E) 53

GO ON TO THE NEXT PAGE ▶▶▶

Model Test 9

46. In figure 7, when $\vec{OA}$ is subtracted from $\vec{OB}$, what will be the length of the resultant vector?

(A) $\sqrt{20}$ (B) $\sqrt{45} - \sqrt{29}$ (C) $\sqrt{12}$

(D) $\sqrt{80}$ (E) $\sqrt{45} + \sqrt{29}$

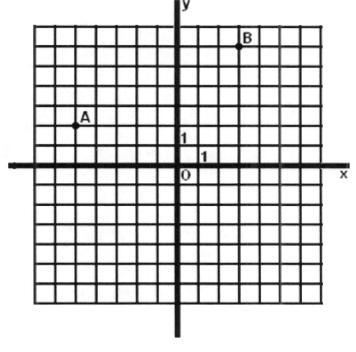

Figure 7

47. Which of the following equations describes the set of all points (x,y) that are equidistant from the x-axis and the point (-3,4)?

(A) $4(y+2)=(x-3)^2$ (B) $4(y-2)=(x+3)^2$ (C) $8(y+2)=(x-3)^2$ (D) $8(y-2)=(x+3)^2$

(E) None of the above.

48. In figure 8, E and F are respectively the midpoints of sides AB and AD of the parallelogram whose lengths are p and q units respectively. If A, G and C are collinear then what is the area of the shaded region?

(A) $\frac{1}{2} \cdot p \cdot q \cdot \sin\theta$ (B) $\frac{3}{4} \cdot p \cdot q \cdot \sin\theta$ (C) $\frac{3}{8} \cdot p \cdot q \cdot \sin\theta$

(D) $p \cdot q \cdot \sin\theta$ (E) $\frac{3}{2} \cdot p \cdot q \cdot \sin\theta$

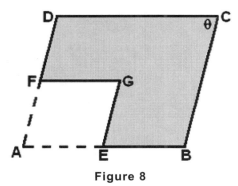

Figure 8

GO ON TO THE NEXT PAGE ▶ ▶ ▶

Model Test 9

49. Which of the following graphs could represent the equation $y=ax^2+bx+c$ where $b^2 - 4ac < 0$?

(A)	(B)	(C)	(D)	(E)

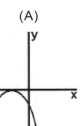 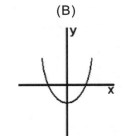 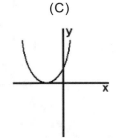 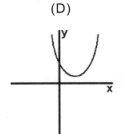 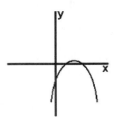

50. "If A is false for some x, then B is false." Which of the following is logically equivalent to the preceding statement?

(A) If B is false, then A is false for some x.

(B) If A is true for all x, then B is true.

(C) If B is true, then A is true for all x.

(D) If B is false, then A is true for all x.

(E) If A is true for all x, then B is false.

S T O P

END OF TEST

(Answers on page 199 – Solutions on page 226)

Model Test 10

Test Duration: 60 Minutes

Directions: For each of the following problems, decide which is the **best** of the choices given. If the exact numerical value is not one of the choices, select the choice that best approximates this value. Then fill in the corresponding oval on the answer sheet.

Notes:

- A calculator will be necessary for answering some (but not all) of the questions in this test. For each question you will have to decide whether or not you should use a calculator. The calculator you use must be at least a scientific calculator; programmable calculators and calculators that can display graphs are permitted.

- For some questions in this test you may have to decide whether your calculator should be in the radian mode or the degree mode.

- Figures that accompany problems in this test are intended to provide information useful in solving the problems. They are drawn as accurately as possible **except** when it is stated in a specific problem that its figure is not drawn to scale.

- All figures lie in a plane unless otherwise indicated.

- Unless otherwise specified, the domain of any function **f** is assumed to be the set of all real numbers **x** for which **f(x)** is a real number.

Reference Information: The following information is for your reference in answering some of the questions in this test.

- Volume of a right circular cone with radius **r** and height **h**: $V = \frac{1}{3}\pi r^2 h$

- Lateral area of a right circular cone with circumference of the base **c** and slant height **l**: $S = \frac{1}{2}cl$

- Volume of a sphere with radius **r**: $V = \frac{4}{3}\pi r^3$

- Surface area of sphere with radius **r**: $S = 4\pi r^2$

- Volume of a pyramid with base area **B** and height **h**: $V = \frac{1}{3}Bh$

1. If $f(a+b) = f(a) \cdot f(b)$ for $a, b \in R$, f(0) can equal

(A) -1   (B) 1   (C) a   (D) b

(E) any real number

2. $f(x) = x + 3x^3 \Rightarrow f(-3) = ?$

(A) -30   (B) -78   (C) 78   (D) -84   (E) 84

GO ON TO THE NEXT PAGE ▶▶▶

Model Test 10

3. Which of the following is not implied by the equation given by $\sqrt{a-b} = c - d$?

(A) a − b is a nonnegative quantity.

(B) d − c is either zero or positive.

(C) a is greater than or equal to b.

(D) d is less than or equal to c.

(E) a equals b and c equals d.

4. 3x + 3y = − 9 and 4y + 4z = 8; x − z = ?

(A) -5   (B) -3   (C) -2   (D) 2   (E) 5

5. If p and q are distinct prime numbers greater than 2, which of the following can be prime?

(A) p^3 (B) pq   (C) 7p   (D) p + q   (E) p − q

6. If the square of the cube root of p is 7, then p can be

(A) − 18.52   (B) − 2.65   (C) 3.66   (D) 49   (E) 343

7. Which of the following restrictions must hold so that the matrix multiplication **AB** can be performed, where the ordered pairs (r_A, c_A) and (r_B, c_B) represent the (number of rows, number of columns) of matrices A and B respectively?

(A) $r_A = c_A$ (B) $r_B = c_B$ (C) $r_A = c_B$ (D) $c_A = r_B$ (E) $r_A \cdot c_A = r_A \cdot c_A$

GO ON TO THE NEXT PAGE ▶▶▶

Model Test 10

8. The average of seven numbers is 10 and six of them are the first six positive even integers. What is the remaining number?

(A) 12   (B) 18   (C) 28   (D) 42   (E) 70

9. For the rectangular box given in figure 1, what is the area of the shaded surface?

(A) 35

(B) 60

(C) 71

(D) 84

(E) 91

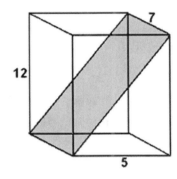

Figure 1

Figure not drawn to scale

10. It is given that (a, b) is a point on f(x) which is then translated to become g(x) where g(x) = f(x + 2) + 1. The point (a, b) becomes

(A) (a + 2, b + 1)   (B) (a − 2, b + 1)   (C) (a + 2, b − 1)   (D) (a − 2, b − 1)   (E) (a + 1, b − 2)

11. What is the range of the function given by $f(x) = \dfrac{4}{|2x - 6| - 2}$?

(A) y < -2 or y > 0   (B) y < 0 or y > 2   (C) All real numbers except 2 and 4

(D) y ≤ -2 or y > 0   (E) -2 < y < 0

GO ON TO THE NEXT PAGE ▶▶▶

Model Test 10

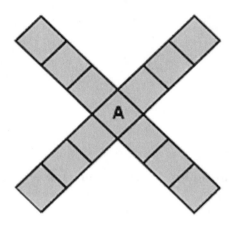

Figure 2

12. Set X contains 13 consecutive integers starting with an odd integer and each number in set X will be placed in a different square in figure 2 above so that the sum of all numbers in the squares from the top left corner to the bottom right corner is to be same as the sum of the numbers in the squares from the bottom left corner to the top right corner. Which of the following must be correct?

(A) A can be no number in set X.

(B) A can be any odd number in set X.

(C) A is the median of the numbers in set X.

(D) A is either the maximum or the minimum of the numbers in set X.

(E) A is either of the maximum, minimum or median of the numbers in set X.

13. Emrah and Defne are talking about the party that Emrah will give. He says to Defne "I am planning that there will be totally 10 people of our class in the party including you and Faye." If there are totally 25 people in Emrah's class, in how many ways can he invite his classmates to the party?

(A) C(25,10)   (B) C(23,8)   (C) C(22,7)   (D) P(23,8)   (E) P(22,7)

14.   m: $2y - 3x = 4$

        n: $y + 5x - 2 = 0$

What is the intersection point of the lines m and n given by the equations above?

(A) (-1, 1)   (B) (1, -2)   (C) (0, 1)   (D) (0, 2)   (E) (2, 0)

GO ON TO THE NEXT PAGE ▶▶▶

Model Test 10

15. Which of the following equations is equivalent to y = 2x?

I. $\ln y = \ln(2x)$

II. $e^y = e^{2x}$

III. $y^4 = 16x^4$

(A) I only   (B) II only   (C) III only   (D) I and II only   (E) I, II and III

16. If (x, y) are the points on the curve given by $y = x^3 - 2x + 1$, what will be the equation of the reflection of this curve about the x axis?

(A) $-x^3 + 2x + 1$ (B) $-x^3 + 2x - 1$ (C) $-x^3 - 2x - 1$

(D) $x^3 - 2x + 1$ (E) $x^3 + 2x + 1$

17. The cost C in dollars of producing n items is given by the function defined as
$C(n) = 13.8n + 4.51$. At most how many dozens of items can be produced with $1500?

(A) 8   (B) 9   (C) 10   (D) 108   (E) 109

18. Each numeral in figure 3 represents a different angle and
$\tan(x) = \cot(y)$; y can be

(A) 3 only

(B) 4 only

(C) 1 and 2 only

(D) 2 and 3 only

(E) 1 and 3 only

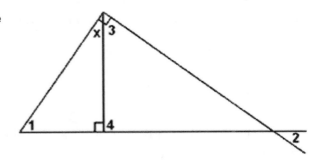

Figure 3

Figure not drawn to scale

GO ON TO THE NEXT PAGE ▶▶▶

Model Test 10

19. If $f(x) = x^3 - 1$ and $g(x) = (x - 1)^3$ then $g(f(2)) - f(g(2))$?

(A) 0   (B) -7   (C) -276   (D) -512   (E) 216

20. It is given that $f(x) = \dfrac{1}{\sin(x)}$; $AB = 2BC = 8$; and $\hat{B}$ is a right angle for the

triangle given in figure 4. What is the value of $f(\hat{A})$?

(A) $\dfrac{1}{\sqrt{5}}$ (B) $\dfrac{2}{\sqrt{5}}$ (C) $\dfrac{\sqrt{5}}{2}$ (D) $\sqrt{5}$ (E) 2

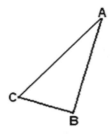

Figure 4

21. The equation given by $y^2 + y = x^2 + x$ represents

(A) A circle   (B) An ellipse   (C) A pair of straight lines   (D) A hyperbola   (E) A parabola

22. Three distinct planes are given such that two of the planes are perpendicular to each other. The set of points that are common to all three planes can consist of

    I. a point.   II. a line.   III. two lines.

(A) I only   (B) II only   (C) I and II only   (D) II and III only   (E) I, II and III

23. It is given that S_1 and S_2 are the lateral surface areas and V_1 and V_2 are the volumes of two similar cylinders 1 and 2 respectively. If $S_1:S_2 = 4:5$ then what is $V_1:V_2$?

(A) $\dfrac{2}{\sqrt{5}}$ (B) $\dfrac{8}{\sqrt{5}}$ (C) $\dfrac{8}{2\sqrt{5}}$ (D) $\dfrac{5\sqrt{5}}{8}$ (E) $\dfrac{8\sqrt{5}}{25}$

GO ON TO THE NEXT PAGE ▶▶▶

Model Test 10

24. What is the greatest negative x intercept of $f(x) = x^3 - 4x - 1$?

(A) -0.254   (B) -1.861   (C) -2.115   (D) 2.115

(E) None of the above

25. If $3x + 1 > 2^x$ then what is the solution set for x?

(A) $0 < x < 3.54$ (B) $0 \leq x \leq 3.54$ (C) $x < 3.54$ (D) $x > 0$ (E) $x < 0$ or $x>3.54$

26. Given in figure 5 is a quadratic function of the form $f(x) = x^2 + mx + n$.
a = ?

(A) (2, 0)   (B) (0, 2)   (C) (1, 0)   (D) (1.5, 0)   (E) (0, 1.5)

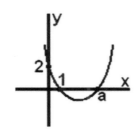

Figure 5

Figure not drawn to scale

27. What are all values for x such that $\sin(2x) = \sin x$ and $0° < x < 360°$?

(A) x = 0°   (B) x = 60°, 300°   (C) x = arccos(0.5)

(D) x = 60°, 180°, 300°   (E) x = 0°, 60°, 180°, 300°

28. If $f(3x - 4) = 4x - 10$ and $f(E) = 2$ then E=?

(A) 2   (B) 3   (C) 5   (D) 9   (E) 12

29. If $\log_2 x^3 = 11$ then x=?

(A) 4.94   (B) 4.95   (C) 12.69   (D) 12.70   (E) 682.67

GO ON TO THE NEXT PAGE ▶▶▶

Model Test 10

30. What are all values of x in the interval $0 \leq x \leq 2\pi$ for which $\sin(x) = c$ and $\cos(x) = \sqrt{1-c^2}$?

(A) $0 \leq x \leq \dfrac{\pi}{2}$ or $\pi \leq x \leq \dfrac{3\pi}{2}$

(B) $0 \leq x \leq \dfrac{\pi}{2}$

(C) $\dfrac{\pi}{2} \leq x \leq \dfrac{3\pi}{2}$

(D) $0 \leq x \leq \dfrac{\pi}{2}$ or $\dfrac{3\pi}{2} \leq x \leq 2\pi$

(E) $0 \leq x \leq 2\pi$

31. There are n lines in a plane which intersect in such a way that each line intersects all others and no intersection point belongs to more than two lines. What is the total number of intersection points in terms of n?

(A) n^2

(B) $n^2 + n$

(C) $n^2 - n$

(D) $\dfrac{n^2 + n}{2}$

(E) $\dfrac{n^2 - n}{2}$

32. Which of the following equations when coupled with the equation $2x - 4y = 3$ gives infinitely many solutions?

(A) $x - 2y = 2$

(B) $6y - 3x = -4.5$

(C) $2y = 1 + x$

(D) $2x = 4y + 5$

(E) $2x + y = 6$

33. In two electronic stores A and B the profit margins are 9.75% and 8.50% respectively. If the price of a certain brand of plasma TV differs by $18.75 in the two stores then what is the wholesale price of the plasma TV assuming that both stores work with the same distributor?

(A) 1300

(B) 1400

(C) 1500

(D) 1600

(E) 1700

GO ON TO THE NEXT PAGE ▶▶▶

Model Test 10

11. Graph of f(x) is given in figure 2. What will be the graph of f(-|x|)?

(A)

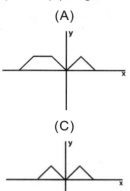

(B)

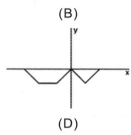

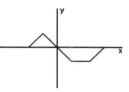

Figure 6

(C)

(D)

(E)

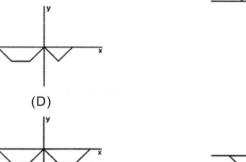

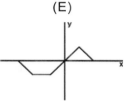

35. Which of the following angle measures does not satisfy the inequality $\sin^2 x > \cos x$?

(A) 6.18 (B) 9.27 (C) 10.4 (D) 99.8° (E) 218°

36. If $\sin 20° = A$ and $\sin x° = -A$, then x can equal

    I. 160°   II. 200°   III. 340°

(A) I only   (B) II only   (C) III only   (D) I and II only   (E) II and III only

37. As is given in figure 7, for what value of p is |AC − BC| a maximum?

(A) 12.0   (B) 11.3   (C) 10.9   (D) 10.3

(E) It cannot be determined from the information given.

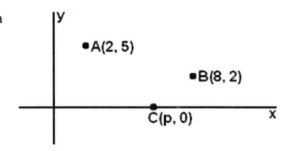

Figure 7

Figure not drawn to scale

GO ON TO THE NEXT PAGE ▶▶▶

(135)

Model Test 10

38. A three dimensional object has 5 vertices and 8 edges. How many faces does it have if each face is in the shape of a convex polygon?

(A) 4   (B) 5   (C) 6   (D) 7   (E) It cannot be determined from the information given.

Figure 8

39. The graph given in figure 8 is the solution set of which of the following inequalities?

(A) $|x - 4| \leq 11$ (B) $|x + 11| < 4$ (C) $|x - 11| < 4$ (D) $|x + 4| < 11$ (E) $|x - 4| < 11$

40. When the statistics of a set of data is calculated, the mean and standard deviation are found to be 10 and 2 respectively. A new data set is obtained by decreasing each data in the previous set by 10 and then dividing the result by 2. Which of the following gives the mean and the standard deviation of the new set?

(A) mean = 1; standard deviation = 0   (B) mean = 0; standard deviation = 1

(C) mean = 10; standard deviation = 2   (D) mean = 10; standard deviation = 1

(E) mean = 0; standard deviation = 2

41. Six students gather in a group everyday and study together for their finals. They make a plan to complete their studies in five days working between 8 AM and 5 PM and each day at noon one person is chosen to buy lunch for all. What is the probability that a different student is chosen on each of the five days?

(A) $\left(\dfrac{1}{6}\right)^4$ (B) $\left(\dfrac{1}{6}\right)^5$ (C) $\dfrac{5 \cdot 4 \cdot 3 \cdot 2}{6 \cdot 6 \cdot 6 \cdot 6}$ (D) $\dfrac{1}{6 \cdot 5 \cdot 4 \cdot 3 \cdot 2}$ (E) $\dfrac{5 \cdot 4 \cdot 3 \cdot 2 \cdot 1}{6 \cdot 6 \cdot 6 \cdot 6 \cdot 6}$

GO ON TO THE NEXT PAGE ▶▶▶

Model Test 10

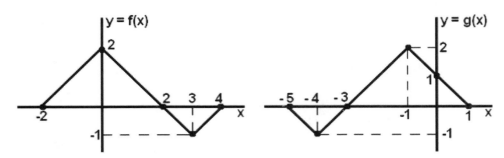

Figure 9

42. If in figure 9 above g(x) = f(ax + b) then (a, b)=?

(A) (-1, 1)　　　　(B) (1, -1)　　　　(C) (-1, -1)　　　　(D) (1, 1)　　　　(E) (-1, 0)

43. Point A(a, b) is reflected across the x axis to give point B. Point B is then reflected across the line y = x to give point C. Finally point C is reflected across the y axis to give point D. What is the distance between the points C and D?

(A) 2a　　　　(B) 2b　　　　(C) |2a|　　　　(D) |2b|　　　　(E) $\sqrt{2 \cdot (a+b)^2}$

44. A tour operator charges every person by \$a when a minimum of n people join an excursion. However for every 5 more people who decide to join the excursion, the operator agrees to decrease price per person by \$b. If 5m more people decide to join the tour what will be the total dollar amount that will be collected by the operator?

(A) (a – b)·(n + m)　　　　　(B) (a – b)·(n + 5m)　　　　　(C) (a – mb)·(n + 5m)

(D) (a + mb)·(n – 5m)　　　　(E) (a + mb)·(n + 5m)

45. What is the amplitude of the function given by f(x) = 3·sin²(2x) + 1?

(A) -3　　　　(B) -1.5　　　　(C) 1.5　　　　(D) 3　　　　(E)4

GO ON TO THE NEXT PAGE ▶▶▶

46. As x approaches zero what happens to the value of $\dfrac{\sin(2x^2)}{x^2}$?

(A) It becomes 0.　　(B) It approaches 2.　(C) It approaches 4.　(D) It becomes undefined.

(E) It approaches a positive quantity whose value cannot be determined.

47. If the Cartesian equation given by $\dfrac{x^2}{4}+\dfrac{y^2}{9}=1$ is parameterized so that x is defined as $2\cos\theta$, then y

can be defined as

　　I. $-3\sin\theta$　　　　II. $3\sin\theta$　　　　III. $3\cos\theta$

(A) I only　　　　(B) II only　　　　(C) I or II only　　　　(D) I or III only　　　　(E) I, II or III.

48. If the points (2, -1, 0), (2, 1, 0), (2, -1, 2) are (0, 1, 2) are four of the vertices of a cube in the three dimensional coordinate system, which of the following point is also a vertex of the same cube?

(A) (2, 2, 2)　　　　(B) (2, -2, 2)　　　　(C) (-2, -1, 0)　　　　(D) (2, 1, 2)　　　　(E) (-1, -2, 0)

49. Which of the following is incorrect for the function defined as $f(x)=\dfrac{x^2}{x^2-1}$?

(A) It has one horizontal and two vertical asymptotes.

(B) Its graph is symmetric with respect to the y axis.

(C) Its domain consists of all real numbers except 1.

(D) Its range is given by $y > 1$ or $y \le 0$.

(E) It has one and only one zero.

50. A sequence is recursively defined by $a_1 = 1$; $a_2 = 2$ and $a_{n+2} = a_{n+1} + a_n$. where n is an integer greater than or equal to 1. How many of the first 10000 terms of this sequence are odd?

(A) 3333　　　　(B) 3334　　　　(C) 6666　　　　(D) 6667　　　　(E) 6789

S T O P

END OF TEST

(Answers on page 199 – Solutions on page 229)

Model Test 11

Test Duration: 60 Minutes

Directions: For each of the following problems, decide which is the **best** of the choices given. If the exact numerical value is not one of the choices, select the choice that best approximates this value. Then fill in the corresponding oval on the answer sheet.

Notes:

- A calculator will be necessary for answering some (but not all) of the questions in this test. For each question you will have to decide whether or not you should use a calculator. The calculator you use must be at least a scientific calculator; programmable calculators and calculators that can display graphs are permitted.

- For some questions in this test you may have to decide whether your calculator should be in the radian mode or the degree mode.

- Figures that accompany problems in this test are intended to provide information useful in solving the problems. They are drawn as accurately as possible **except** when it is stated in a specific problem that its figure is not drawn to scale.

- All figures lie in a plane unless otherwise indicated.

- Unless otherwise specified, the domain of any function **f** is assumed to be the set of all real numbers **x** for which **f(x)** is a real number.

Reference Information: The following information is for your reference in answering some of the questions in this test.

- Volume of a right circular cone with radius **r** and height **h**: $V = \frac{1}{3}\pi r^2 h$

- Lateral area of a right circular cone with circumference of the base **c** and slant height **l**: $S = \frac{1}{2}cl$

- Volume of a sphere with radius **r**: $V = \frac{4}{3}\pi r^3$

- Surface area of sphere with radius **r**: $S = 4\pi r^2$

- Volume of a pyramid with base area **B** and height **h**: $V = \frac{1}{3}Bh$

1. If $\dfrac{3x + 5y}{2} = \dfrac{-m}{4}$ then m=?

(A) $3x + 5y$ (B) $-3x - 5y$ (C) $6x + 10y$ (D) $-6x - 10y$

(E) None of the above

GO ON TO THE NEXT PAGE ▶▶▶

Model Test 11

2. $\dfrac{1}{1+\dfrac{1}{1+x}} = ?$

(A) $\dfrac{x}{x+1}$ (B) $\dfrac{x}{2x+1}$ (C) $\dfrac{x+1}{x+2}$ (D) $\dfrac{x}{x+2}$ (E) $\dfrac{2x}{x+2}$

3. If $-3^{2x}=1$ then x=?

(A) -0.5 (B) 0.5 (C) -1/9 (D) 0 (E) There is no such value of x.

4. If A is a point in the first quadrant and B is a point in the second quadrant then midpoint of AB line segment cannot be

(A) (1, 4) (B) (2, 0) (C) (0, 2.1) (D) $(\dfrac{2}{3}, 8)$ (E) (-3, 6)

5. If an electric saw can cut a long piece of wood into 10 pieces in 90 seconds, then how many seconds are needed to obtain 20 pieces? (Assume that cuts are identical and identical pieces are obtained.)

(A) 90 (B) 100 (C) 180 (D) 190 (E) 200

6. If f(x) is given by f(x)= $\begin{cases} 3x^2 & x \le 1 \\ 4 & x > 1 \end{cases}$, then what is f(x-1)?

| (A) | (B) | (C) | (D) | (E) |

(A) $\begin{cases} 3(x-1)^2 & x \le 1 \\ 4 & x > 1 \end{cases}$ (B) $\begin{cases} 3x^2 & x \le 2 \\ 4 & x > 2 \end{cases}$ (C) $\begin{cases} 3(x-1)^2 & x \le 2 \\ 4 & x > 2 \end{cases}$ (D) $\begin{cases} 3(x-1)^2 & x \le 0 \\ 4 & x > 0 \end{cases}$ (E) $\begin{cases} 3(x-1)^2 & x \le -1 \\ 4 & x > -1 \end{cases}$

GO ON TO THE NEXT PAGE ►►►

Model Test 11

7. A number x divided by the square of its reciprocal results in another number which is less than x. Which of the following can be x?

(A) $-\dfrac{9}{7}$ (B) -0.1   (C) 1   (D) 1.67   (E) There is no such number.

8. If $\dfrac{9}{x+1}=0$ then x-1 is

(A) -2   (B) -1   (C) 0   (D) 1   (E) Undefined.

9. If $x^2+(y+2)^2=0$ then y is

(A) 0   (B) 2   (C) -2x   (D) x-2   (E) There is not enough information to find y.

10. If 2x-2y=4 and 3x+3y= -9 then x^2-y^2=?

(A)3   (B) 6   (C) 2   (D) -3   (E) -6

11. SAT II Math 2C tests ready for packaging at *RUSH* Publications come down from the printing house in a single file. Examiner A checks every third test, beginning with the third. Examiner B checks every fifth test, beginning with the fifth. If 1000 tests came down the printing house while both inspectors were working on Monday, how many of these tests were not checked by either of these two examiners?

(A) 533   (B) 467   (C) 267   (D) 200   (E) 66

GO ON TO THE NEXT PAGE ▶▶▶

Model Test 11

12. E and F are respectively the midpoints of sides AD and AB in the parallelogram ABCD shown in figure 1. The area of pentagon BCDEF is

(A) 12   (B) 18   (C) 24

(D) 32   (E) 42

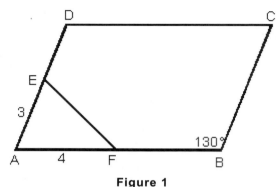

Figure 1
Figure not drawn to scale

13. If a items are bought for $b, and b items are sold for $a, given that a and b are both integers where a>b, what is the rate of the profit made?

(A) $\left(\dfrac{b}{a}-\dfrac{a}{b}\right)\dfrac{100b}{a}\%$ (B) $\left(\dfrac{a}{b}-\dfrac{b}{a}\right)100\%$ (C) $\left(\dfrac{a}{b}-\dfrac{b}{a}\right)\dfrac{100a}{b}\%$ (D) $\left(\dfrac{b}{a}-\dfrac{a}{b}\right)100\%$ (E) $\left(\dfrac{a}{b}-\dfrac{b}{a}\right)\%$

14. Sets A and B are given as follows: A = {1, 2, 3, 4} and B = {a, b, c, d}. How many of the following relations are functions from A to B?

    I. {(1, a), (2, a), (3, a), (4, a)}   II. {(1, a), (1, b), (2, b), (3, b), (4, c)}

    III. {(a, 1), (b, 2), (c, 3), (d, 4)}   IV. {(1, d), (2, c), (3, a)}

    V. {(1, b), (2, a), (3, c), (4, c)}

(A) 1   (B) 2   (C) 3   (D) 4   (E) 5

15. If $f(x)=ax^2+bx+c$ where $ac<0$ then which of the following can be the graph of $f(x)$?

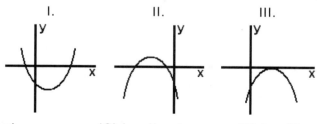

(A) I only   (B) II only   (C) I or II only   (D) I or III only   (E) I, II or III

GO ON TO THE NEXT PAGE ▶▶▶

Model Test 11

16. From 10 experts and 20 trainees, how many different teams can be made such that each team consists of an expert and two trainees?

(A) 4000   (B) 3800   (C) 2000   (D) 1900

(E) None of the above

17. Each of the consecutive even integers between 1 and 1000 inclusive will be written on a paper exactly once. How many digits have to be used?

(A) 499   (B) 998   (C) 1000   (D) 500   (E) 1448

18. What is the identity element for the operation $x@y=x+y-2xy$

(A) -1   (B) -0.5   (C) 0   (D) 0.5   (E) 1

19. A quadratic equation with integral coefficients has distinct rational roots. Which of the following must be correct?

(A) Its discriminant is negative.

(B) Its discriminant is nonnegative.

(C) Its discriminant is positive definite.

(D) Its discriminant is a positive perfect square.

(E) Its discriminant is a nonnegative perfect square.

20. Roses are to be planted inside a semicircular flower garden with a diameter of 20 feet such that for every 15 square inches one rose will be planted. If roses are sold in packs of 8 and each pack is worth 7 dollars, what will be the cost of flowering the whole garden to the nearest hundred dollars?

(A) 500   (B) 600   (C) 700   (D) 1300   (E) 2600

GO ON TO THE NEXT PAGE ▶▶▶

Model Test 11

21.   N = 1234567891011 ... 979899

The integer N is formed by writing the positive integers in a line one after another, starting with 1 and ending with 99, as shown above. Counting from left, what is the 81st digit of N?

(A) 3　　　　　　(B) 4　　　　　　(C) 5　　　　　　(D) 6　　　　　　(E) 8

22. The front, side, and bottom faces of a rectangular solid have areas of 15 square inches, 5 square inches, and 3 square inches, respectively. What is the volume of the solid?

(A) 15 cubic inches　　　(B) 60 cubic inches　　　(C) 120 cubic inches　　　(D) 225 cubic inches

(E) It cannot be determined from the information given.

For questions 23 through 25 please refer to the following table that shows the preferences of the engineering students by major at the beginning and the end of the sophomore year of the university.

		Major at the end of the sophomore year					
		Civil	Electrical	Computer	Environmental	Industrial	Totals at the beginning of the year
Major at the beginning of the sophomore year	Civil	45	5	8	2	6	66
	Electrical	6	40	5	0	17	68
	Computer	5	20	55	10	15	105
	Environmental	2	5	4	25	5	41
	Industrial	10	7	4	2	70	93
	Totals at the end of the year	68	77	76	47	113	

23. How many students who chose to major in civil engineering at the beginning of the sophomore year changed their major to electrical engineering at the end of the sophomore year?

(A) 2　　　　　　(B) 5　　　　　　(C) 6　　　　　　(D) 7　　　　　　(E) 8

24. By the end of the sophomore year the greatest drop out in enrollment was in

(A) Civil engineering　　　　　　(B) Electrical engineering　　　　　　(C) Computer engineering

(D) Environmental engineering　　　　　　(E) Industrial engineering

GO ON TO THE NEXT PAGE ▶▶▶

Model Test 11

25. How did the enrollment to all engineering branches change at the end of the sophomore year compared to the beginning of the sophomore year?

(A) Decreased by 8   (B) Increased by 8   (C) Did not change

(D) Decreased by 6   (E) Increased by 6

26. $x^4+x^2+1=0$ $x^4 + \dfrac{1}{x^4} = ?$

(A) 0   (B) -1   (C) 1   (D) -2

(E) None of the above.

27. If the terms of a geometric sequence are denoted as g_n and $g_3 = 32$ and $g_6 = 4$; then what is $g_{11} + g_{15}$?

(A) .06   (B) .25   (C) $\dfrac{17}{32}$ (D) $\dfrac{17}{64}$ (E) $\dfrac{17}{128}$

28. A square has two of its vertices at (1, 1) and (1, -1). How many such squares can there be?

(A) 1   (B) 2   (C) 3   (D) 4   (E) 5

29. The graph given in figure 2 corresponds to which of the following relations?

(A) |x| + |y| = A   (B) |x| - |y| = A   (C) |y| - |x| = A

(D) x + |y| = A   (E) |x| + y = A

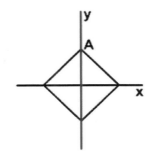

Figure 2

GO ON TO THE NEXT PAGE ▶ ▶ ▶

Model Test 11

30. If θ is a positive obtuse angle then $2\sin\theta\cos\theta-(\sin\theta+\cos\theta)^2$=?

(A) 1 (B) -1 (C) 0 (D) 2 (E) -2

31. The points that satisfy the inequality $\dfrac{x}{y}<1$ lie in which quadrants?

(A) I, and II (B) III and IV (C) I, II and III (D) I, III and IV (E) All four quadrants

32. Number of real roots of a 7th degree real polynomial cannot be

(A) 0 (B) 1 (C) 3 (D) 5 (E) 7

33. Two pegs A and B are nailed on a rectangular wall 27ft long and 13ft tall in the following way: Peg A is nailed 3 ft down from the upper edge and 5ft left from the right edge of the wall. Peg B is nailed 2 ft up from the bottom edge and 7ft right from the left edge of the wall. What is shortest distance in feet between these pegs?

(A) 8 (B) 15 (C) 17 (D) 23

(E) None of the above

34. For the cube given in figure 3 AB and BC are the diagonals of the corresponding square faces. If measure of $\angle$ABC is α, then

(A) 30° < α < 45° (B) 45° < α < 60° (C) 55° < α < 65°

(D) 60° < α < 70° (E) 75° < α < 90°

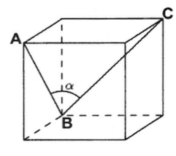

Figure 3

GO ON TO THE NEXT PAGE ▶ ▶ ▶

35. In figure 4, shaded pentagonal region S has an area of 720 square inches. What is the area in square feet of the region that consists of the points (x, -2y) where (x,y) is in S?

(A) 1440   (B) 720   (C) 10   (D) 5

(E) None of the above

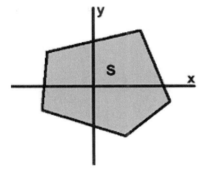

Figure 4

36. l_1: 2x–my+3=0

       l_2: 3x+2y+n=0

If the coplanar lines given by l_1 and l_2 have no intersection points what are all possible values for m and n?

(A) n = -4/3 and m ≠ 9/2   (B) m ≠ -4/3

(C) m = -4/3 and n = 9/2   (D) m = -4/3

(E) m = -4/3 and n ≠ 9/2

37. If $f(x) = \dfrac{mx+n}{px-q}$ then $f^{-1}(x)=$?

(A) $\dfrac{mx+n}{px-q}$ (B) $\dfrac{qx+n}{px-m}$ (C) $-\dfrac{mx+n}{px-q}$ (D) $\dfrac{qx+n}{px+m}$ (E) $\dfrac{px-q}{mx+n}$

38. The relation given by $\dfrac{x^2}{25} - \dfrac{y^2}{144} = 1$ has how many of the following types of symmetry?

   I. Symmetry in the origin   II. Symmetry in the x axis

   III. Symmetry in the y axis   IV. Symmetry in the line y=x.

   V. Symmetry in the line y=-x

(A) 1   (B) 2   (C) 3   (D) 4   (E) 5

GO ON TO THE NEXT PAGE ▶▶▶

Model Test 11

39. If the parabola given by $x^2 - 2x + y = 0$ is tangent to the line given by $y=mx+9$ then the negative value of m is

(A) 8   (B) 4   (C) 0   (D) -4   (E) -8

1. Shift right for 3 units
2. Shift down for 2 units
3. Reflect in the y axis

40. If the sequence of transformations above is to be applied, in order, to $f(x) = x^2$ such that the result of each transformation is used as the starting expression for the next transformation, which of the following represents the outcome after step 3 has been completed?

(A) $(x-3)^2 + 2$ (B) $(x-3)^2 - 2$ (C) $(x+3)^2 - 2$ (D) $(x+3)^2 + 2$ (E) $-(x+3)^2 - 2$

41. In figure 5, a semicircle is given. If P is the center and R is the radius then what is the shortest distance from point C to segment AB in terms of R and α?

(A) $2R\cos2\alpha$ (B) $2R\sin(2\alpha)$ (C) $R\sin\alpha\cdot\cos\alpha$

(D) $2R\sin\alpha\cdot\cos\alpha$ (E) $4R\sin\alpha\cdot\cos\alpha$

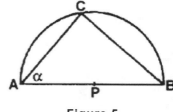

Figure 5

42. If 8 identical iron spheres are to be melted into a big sphere; how will the surface area of the resulting sphere compare to the total surface area of the small spheres?

(A) It will be twice the total surface area of the small spheres.

(B) It will be the same as the total surface area of the small spheres.

(C) It will be one fourth of the total surface area of the small spheres.

(D) It will be half as much as the total surface area of the small spheres.

(E) It will be 4 times as much as the total surface area of the small spheres.

GO ON TO THE NEXT PAGE ▶▶▶

Model Test 11

43. Given that $\ln(y^2)=x$, what is y in terms of x?

(A) $\dfrac{1}{2}e^x$ (B) $2e^x$ (C) $e^{\frac{x}{2}}$ (D) $\pm e^{\frac{x}{2}}$ (E) $e^{\pm\frac{x}{2}}$

44. If a and b are positive integers such that $\dfrac{6}{7}<\dfrac{a}{b}<\dfrac{7}{8}$ then what is the smallest value that b can take?

(A) 13   (B) 15   (C) 43   (D) 56   (E) 112

45. If $i=\sqrt{-1}$ then $(i-1)^2(2i+1)=$?

(A) 4+2i   (B) 4-2i   (C) 2-4i   (D) 2+4i   (E) 2i-4

46. Cem is a genius student whose mathematics teacher made the following claim: "If Cem takes the Math Level 2 Subject Test, he will definitely score 800." However it turned out that Cem did not score 800 in the Math Level 2 Subject Test. Assuming that the claim made by Cem's teacher was correct, which of the following can be deduced?

(A) Cem took the test and some of his answers were incorrect.

(B) Nobody was able to score 800 on that test day, neither did Cem.

(C) Cem left some questions unanswered in the test.

(D) Some of Cem's friends scored 800 on the test.

(E) Cem did not take the test.

47. Kerem has some coins worth of 195 cents that are made up of quarters and nickels only. Which of the following cannot be the total number of coins he has?

(A) 7   (B) 11   (C) 15   (D) 19   (E) 23

GO ON TO THE NEXT PAGE ▶▶▶

Model Test 11

48. How many terms are there in the expansion of $(x-2y)^{-4}$?

(A) 2   (B) 3   (C) 4   (D) 5   (E) More than 5

49. If $A_{2\times4}$ and $B_{4\times2}$ are matrices, then what will be the dimensions of BA?

(A) 2x4   (B) 4x2   (C) 2x2   (D) 4x4

(E) None of the above

50. Which of the following is false regarding the properties of the real polynomial functions whose graphs are given below if each function has no local maximum or local minimum points that are not visible?

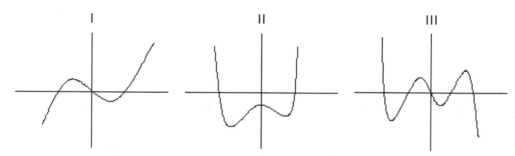

(A) The range of polynomials given by I and III are all real numbers.

(B) The polynomial given by II has an even degree.

(C) The polynomial given by I has exactly 3 zeros.

(D) The least degree of the polynomial given by III is 5.

(E) The polynomial given by II has at least two complex zeros.

S T O P

END OF TEST

(Answers on page 199 – Solutions on page 231)

Model Test 12

Test Duration: 60 Minutes

Directions: For each of the following problems, decide which is the **best** of the choices given. If the exact numerical value is not one of the choices, select the choice that best approximates this value. Then fill in the corresponding oval on the answer sheet.

Notes:

- A calculator will be necessary for answering some (but not all) of the questions in this test. For each question you will have to decide whether or not you should use a calculator. The calculator you use must be at least a scientific calculator; programmable calculators and calculators that can display graphs are permitted.

- For some questions in this test you may have to decide whether your calculator should be in the radian mode or the degree mode.

- Figures that accompany problems in this test are intended to provide information useful in solving the problems. They are drawn as accurately as possible **except** when it is stated in a specific problem that its figure is not drawn to scale.

- All figures lie in a plane unless otherwise indicated.

- Unless otherwise specified, the domain of any function **f** is assumed to be the set of all real numbers **x** for which **f(x)** is a real number.

Reference Information: The following information is for your reference in answering some of the questions in this test.

- Volume of a right circular cone with radius **r** and height **h**: $V = \frac{1}{3}\pi r^2 h$

- Lateral area of a right circular cone with circumference of the base **c** and slant height **l**: $S = \frac{1}{2}cl$

- Volume of a sphere with radius **r**: $V = \frac{4}{3}\pi r^3$

- Surface area of sphere with radius **r**: $S = 4\pi r^2$

- Volume of a pyramid with base area **B** and height **h**: $V = \frac{1}{3}Bh$

1. $3a = b$; $c = \frac{b}{4}$; $\frac{a}{c} = ?$

(A) $\frac{3}{4}$ (B) $\frac{4}{3}$ (C) $\frac{1}{12}$ (D) $\frac{1}{4}$ (E) $\frac{1}{3}$

2. Which of the following is the least for the same positive integer value of x?

(A) $\frac{2}{3x}$ (B) $\frac{3x}{2}$ (C) $\frac{2}{3x+1}$ (D) $\frac{2}{3x-1}$ (E) $\frac{3x+1}{2}$

GO ON TO THE NEXT PAGE ▶▶▶

Model Test 12

3. If $\dfrac{x}{2}+0.1x-\dfrac{15x}{100}=18$ then x=?

(A) -20   (B) -1.5   (C) 8.1   (D) 24   (E) 40

4. If $g(x, y, z) = (x^2, y^3, z)$ and $h(x, y, z) = (2x, 3y, -2z)$; then $g(h(a, b, c)) =$

(A) $(4a^2, 27b^3, -2c^2)$ (B) $(4a^2, 27b^3, -2c)$ (C) $(2a^2, 3b^3, -2c)$ (D) $(2a, 3b, -2c)$ (E) $(8a, 27b, -2c)$

5. $f(x) = 2^x - 3^{x-1}$; $f(2t) = 1$; t can be

(A) -1   (B) -0.5   (C) 0   (D) 1   (E) 1.1

6. $f(x) = 3\sqrt{x} + \sqrt{3x}$; $3 \cdot f(3) = ?$

(A) $3\sqrt{3}+3$ (B) $9\sqrt{3}+1$ (C) $3 \cdot (\sqrt{3}+3)$ (D) $3 \cdot (\sqrt{3}+3)$ (E) $9 \cdot (\sqrt{3}+1)$

7. If the points A(7, 24) and B(-6, 8) are the endpoints of a diameter of circle C, what are the coordinates of the center of circle C?

(A) (1, 32)   (B) (13, 16)   (C) (0.5, 16)   (D) (25,10)

(E) None of the above

GO ON TO THE NEXT PAGE ▶▶▶

Model Test 12

8. If x and y are real numbers, and if $(x + 2i)^2 = y - 6i$, then y =

(A) -1.75 (B) -1.5 (C) -1.5 (D) 1.75 (E) 3

9. If $\cos x = \dfrac{3}{a}$ and $-\dfrac{\pi}{2} < x < \dfrac{\pi}{2}$ then sinx=?

(A) $\pm\dfrac{\sqrt{a^2-9}}{3}$ (B) $\dfrac{\sqrt{a^2-9}}{3}$ (C) $\pm\dfrac{\sqrt{a^2-9}}{a}$ (D) $\dfrac{a^2-9}{a^2}$ (E) $\pm\dfrac{\sqrt{9-a^2}}{a}$

10. What number must be added to each of the numbers 3, 15 and 51 so that they form a consecutive terms of a geometric sequence?

(A) 3 (B) 4 (C) 5 (D) 6 (E) 9

11. Based on the information given in the table for g(x) and h(x), g(h(2)) = ?

(A) -2 (B) -1 (C) 0 (D) 1 (E) 3

x	g(x)	x	h(x)
1	0	1	-3
2	3	2	3
3	-1	3	-2
4	2	4	1

12. What is the range of $|\sin(2x)|$?

(A) $0 < y < 1$ (B) $0 \le y \le 1$ (C) $0 \le y \le 2$ (D) $-1 \le y \le 1$ (E) $0 \le y \le 0.5$

13. At the end of 2000, the number of residents in the city of Mercuria was 133,000. If the population has been decreasing by exactly 3% per year, what will be the population of city Mercuria at the end of 2010 rounded to the nearest 100 people?

(A) 93,100 (B) 98,000 (C) 98,100 (D) 101,100 (E) 178,700

GO ON TO THE NEXT PAGE ▶▶▶

Model Test 12

14. Point (-3, 2) is located on a circle in the coordinate plane given by the equation $(x - 1)^2 + (y - 1)^2 = r^2$. If r is the radius of the circle, which of the following points must also lie on the circle?

(A) (1, 2)   (B) (4, 1)   (C) (1, -1)   (D) (5, 0)   (E) (-1, -1)

15. If p is a positive even integer for which 3p + 1 is not a prime number, then what is the minimum value of p?

(A) 0   (B) 1   (C) 4   (D) 6   (E) 8

16. If $\cos\theta = 0.35$ in figure 1 then what is the value of $\dfrac{z}{y}$?

(A) .37   (B) .94   (C) 1.87   (D) 2.68   (E) 2.81

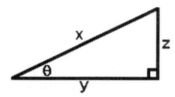

Figure 1
Figure not drawn to scale

17. If $(x+2)(x^4+x^2+1) = 0$ then $(x-2)(x^3+x^2+1) = ?$

(A) -44   (B) -12   (C) 0   (D) 12   (E) 44

18. If f(x) = 4x − 1 and g(x) = 2x + 1 then what is g(x) in terms of f(x)?

(A) g(x) = 2·f(x) + 3   (B) g(x) = 2·f(x) − 3   (C) g(x) = 3·f(x) − 2   (D) $g(x) = \dfrac{f(x)+2}{3}$ (E) $g(x) = \dfrac{f(x)+3}{2}$

GO ON TO THE NEXT PAGE ►►►

Model Test 12

19. If $\dfrac{1}{2x} + \dfrac{1}{4x} - \dfrac{1}{8x} = \dfrac{1}{4}$ then $x^{-1} = ?$

(A) 0.4   (B) 0.285   (C) 0.286   (D) 0.625   (E) 2.5

20. If $f(x) = 3x - 2$ and $-2 < f(x) \leq 4$ then what is the domain of $f(x)$?

(A) $-2 < x \leq 0$ (B) $-2 < x < 0$ (C) $-2 < x \leq 2$ (D) $0 < x \leq 2$ (E) $0 < x < 2$

21. The time t in hours it takes to produce a certain type of machinery of m kilograms is given by the formula $t = 19.6 \cdot m^{\frac{1}{3}}$. How many days are needed to produce a machine that is 4500 kg?

(A) 13   (B) 14   (C) 323   (D) 324   (E) 1225

22. A bag contains five red marbles and four green marbles. A green marble is replaced by a red marble and three marbles are randomly drawn from the bag. What is the probability that all three will be red?

(A) 1 / 12   (B) 1 / 84   (C) 5 / 21   (D) 5 / 42   (E) 40 / 243

23. Under what conditions does x + y + b = 0 represent a horizontal line?

(A) if $0 \leq b \leq 1$ (B) if b = 0   (C) if b < 1   (D) for all b   (E) for no b

GO ON TO THE NEXT PAGE ▶▶▶

Model Test 12

24. If the graph in figure 5 corresponds to $y = |f(-x)|$, then which of the following graphs cannot be the graph of f(x)?

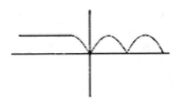

Figure 2

(A) 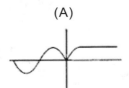   (B) 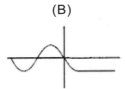   ((C) 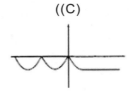   (D) 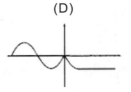   (E)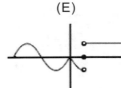

25. There are 5 green, 6 blue and 7 red marbles in a bag. Three marbles are selected randomly from the bag. What is the probability that two of them will be red and one of them is blue?

(A) $\dfrac{7}{136}$ (B) $\dfrac{21}{136}$ (C) $\dfrac{35}{272}$ (D) $\dfrac{35}{816}$

(E) None of the above

26. It is given that $f(x) = 2x - \dfrac{B}{2x}$. If $f(-1) = 4 \cdot f(1)$ then B=?

(A) 2   (B) 3   (C) 4   (D) 5   (E) 6

27. If y is a quadratic function of x then p=?
(A) 29   (B) 35   (C) 39   (D) 45
(E) It cannot be determined from the information given

x	y
1	4
2	7
4	19
6	p

GO ON TO THE NEXT PAGE ▶▶▶

Model Test 12

28. (tanx-secx)(tanx+secx)=?

(A) -1   (B) 1   (C) sinx+tanx   (D) cosx+tanx

(E) None of the above

29. Based on the information given in figure 5, x=?

(A) 22.93   (B) 23.34   (C) 26.78   (D) 26.77   (E) 29.21

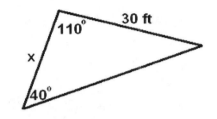

Figure 3

30. As x becomes infinitely large, which of the following functions approaches to 3?

(A) $\dfrac{x+3}{x+1}$ (B) $\dfrac{x+1}{3x-1}$ (C) $\dfrac{-3x+1}{x-1}$ (D) $\dfrac{-3x+1}{-x+1}$ (E) $\dfrac{3x-1}{-x-3}$

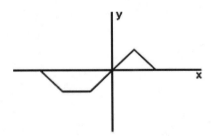

Figure 4

31. Graph of f(x) is given in figure 4 above. What will be the graph of f $^{-1}$(x)?

(A)   (B)   ((C)   (D)   (E)

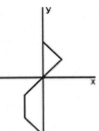 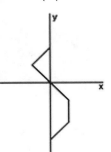 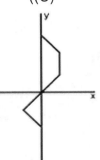 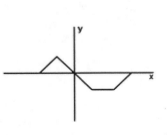 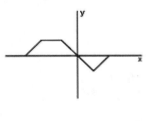

GO ON TO THE NEXT PAGE ►►►

Model Test 12

32. Which of the following sets of numbers has the least standard deviation?

(A) 1005, 1007, 1009   (B) 100, 101, 102   (C) 10000, 10000, 10000   (D) 0.01, 0.02, 0.03   (E) 5, 9, 17

33. The final mathematics scores of the senior class in *RUSH* Academy are given in the graph in figure 6 above. What is the mean score in this class?

(A) 54   (B) 54.8

(C) 55   (D) 60

(E) None of the above

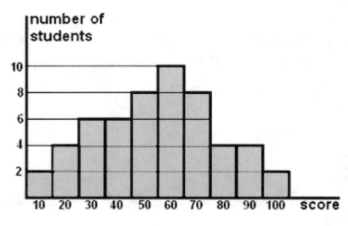

Figure 5

34. It is given that f(x) is a linear function such that $f(x+1) - f(x) = 6$. What is the value of $f^{-1}(x+1) - f^{-1}(x)$?

(A) 1/6   (B) 1/3   (C) 1   (D) 3   (E) 6

35. The function given by $\log_5(5-x)$ intersects the x axis when x=?

(A) 0   (B) 1   (C) 2   (D) 3   (E) 4

36. $f(x) = e^{2x}$. $f^{-1}(4) = ?$

(A) 2981   (B) 0.301   (C) 0.693   (D) 1.204   (E) 2.773

GO ON TO THE NEXT PAGE ▶▶▶

Model Test 12

37. In each of the following cases, solutions contain water and alcohol only:

Case 1: x liters of % 32.5 alcohol solution and y liters of % 27.5 alcohol solution are mixed to make a % 30 alcohol solution.

Case 2: x liters of % 32.5 alcohol solution and y liters of % 25.5 alcohol solution are mixed to make a % 30 alcohol solution.

Case 3: x liters of % 32.5 alcohol solution and y liters of % 25.5 alcohol solution are mixed to make a % 27.5 alcohol solution.

Which one(s) of the following are correct?

I. In Case 1, x = y.

II. In Case 2, x < y.

III. In Case 3, x < y.

(A) I only   (B) II only   (C) III only   (D) I and II only   (E) I and III only

38. f is a function. Suppose the following two statements are true for f:

I. f has a zero at x = 3

II. The graph of $y = f(x)$ has exactly two asymptotes, at x = -1 and y = 2

The function f can be

(A) $f(x) = \dfrac{x-3}{x+1}$ (B) $f(x) = \dfrac{2x}{x+1}$ (C) $f(x) = \dfrac{2x}{x-3}$ (D) $f(x) = \dfrac{x-3}{2x}$ (E) $f(x) = \dfrac{2x-6}{x+1}$

39.   A: 2x + 3y + 4z = 5

        B: 6x + 9y + 12z = 10

        C: − 4x − 6y − 8z = 5

Which of the following are correct about the three dimensional figures represented by A, B and C?

(A) A, B, and C represent parallel lines where B is between A and C.

(B) A, B, and C represent parallel lines where A is between B and C.

(C) A, B, and C represent parallel lines where C is between A and B.

(D) A, B, and C represent parallel planes where B is between A and C.

(E) A, B, and C represent parallel planes where C is between A and B.

GO ON TO THE NEXT PAGE ►►►

Model Test 12

40. Which of the following is a zero of $f(x) = 2x^2 - 4x - 5$?

(A) -2.87   (B) 2.87   (C) 0.87   (D) 1.87   (E) -1.87

41. If the period of $\cos(x)$ is T then which of the following has a period of $\dfrac{T}{3}$?

(A) $\dfrac{\cos(x)}{3}$ (B) $\cos\left(\dfrac{x}{3}\right)$ (C) $\cos\left(x + \dfrac{1}{3}\right)$ (D) $\cos(-3x)$ (E) $3\cos(x)$

42. If $a = b^x$ then $\dfrac{b}{a} = ?$

(A) b^{1-x} (B) b^{x-1} (C) b^x (D) a^{1-x} (E) a^{x-1}

43. Twelve seniors and eight juniors have been nominated to serve on a student board. If the board must consist of six seniors and four juniors, how many different groups of students could serve on the board?

(A) $\dfrac{12!}{6!} \cdot \dfrac{8!}{4!}$ (B) $\dfrac{12!}{6!} + \dfrac{8!}{4!}$ (C) $\dbinom{12}{6} \cdot \dbinom{8}{4}$ (D) $\dfrac{12!}{6! \cdot 6!} + \dfrac{8!}{4! \cdot 4!}$

(E) None of the above

44.   I. $\sin^2 x + 2\cos^2 x = 1$

     II. $3\sin^2 x + 4\cos^2 x = 2$

     III. $4\sin^2 x + 3\cos^2 x = 5$

Which of the above equations have the same solution set for x?

(A) I and II   (B) I and III   (C) II and III

(D) Each equation has a different solution set.

(E) All three equations have the same solution set.

GO ON TO THE NEXT PAGE ▶▶▶

Model Test 12

45. A function is given by $y = 5 \cdot e^{(-0.3x)} \cdot \sin(x) + 2.4$. What is the maximum value of x for which $|y| = 2.3$?

(A) 3.19   (B) 6.16   (C) 9.81   (D) 11.80   (E) There is no such value of x.

46. The height h(t) in meters of a cannon ball thrown vertically upwards from level ground is a function of time given by the formula $h(t) = v_o t - 5t^2$ where v_o is the initial upward velocity of the cannon ball in meters per seconds and t is time in seconds. The cannon ball is thrown upwards with an initial velocity of 20 m/s. How many seconds after the ball reaches its maximum height will the ball be at a height of 15 m?

(A) 1   (B) 2   (C) 3   (D) 4

(E) None of the above

47. Ice-cream is leaking out of a right circular ice-cream cone with a diameter of 8 and height of 16 inches, as shown in figure 9. What volume of the ice cream in cubic inches will remain in the cone when the height AB of the ice cream is 6 inches?

(A) 42.41   (B) 28.27   (C) 14.14

(D) 14.13   (E) 14.10

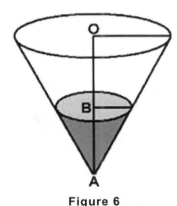

Figure 6
Figure not drawn to scale

48. Which of the following are equal to 1 when x is an obtuse angle?

   I. $\tan(x) \cdot \cot(x)$

   II. $\sec^2(x) - \tan^2(x)$

   III. $\sec(x) \cdot \csc(x)$

(A) I only   (B) II only   (C) I and II only   D) I and III only   (E) I, II and III

GO ON TO THE NEXT PAGE ▶▶▶

Model Test 12

49. A cylinder of infinite height with a radius of 3 inches is

(A) the locus of points equidistant from the surface of a cylinder of infinite height with a radius of 3 inches.

(B) the locus of points at a distance of 3 inches from a given point in a plane.

(C) the locus of points at a distance of 3 inches from a given line in a plane.

(D) the locus of points at a distance of 3 inches from a given point in space.

(E) the locus of points at a distance of 3 inches from a given line in space.

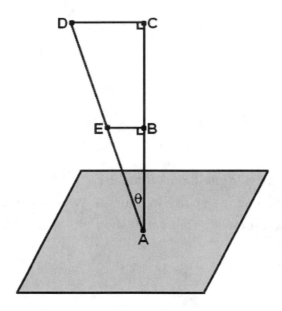

Figure 7

50. As indicated in figure 7 above, the intensity of light at point A on the shaded plane created by a light source at point E is directly proportional with the cosine of the angle θ that the light source makes with the normal to the surface and inversely proportional with the square of the distance between the light source and the point E. If DE = EA = 4EB then how does the intensity of light at point A change if the light source is moved from point E to point D?

(A) Intensity is divided by a factor of 2.   (B) Intensity is divided by a factor of 4.

(C) Intensity is divided by a factor of 2.5.   (D) Intensity is multiplied by a factor of 4.

(E) Intensity is multiplied by a factor of 2.

S T O P

END OF TEST

(Answers on page 199 – Solutions on page 234)

Model Test 13

Test Duration: 60 Minutes

Directions: For each of the following problems, decide which is the **best** of the choices given. If the exact numerical value is not one of the choices, select the choice that best approximates this value. Then fill in the corresponding oval on the answer sheet.

Notes:

- A calculator will be necessary for answering some (but not all) of the questions in this test. For each question you will have to decide whether or not you should use a calculator. The calculator you use must be at least a scientific calculator; programmable calculators and calculators that can display graphs are permitted.

- For some questions in this test you may have to decide whether your calculator should be in the radian mode or the degree mode.

- Figures that accompany problems in this test are intended to provide information useful in solving the problems. They are drawn as accurately as possible **except** when it is stated in a specific problem that its figure is not drawn to scale.

- All figures lie in a plane unless otherwise indicated.

- Unless otherwise specified, the domain of any function **f** is assumed to be the set of all real numbers **x** for which **f(x)** is a real number.

Reference Information: The following information is for your reference in answering some of the questions in this test.

- Volume of a right circular cone with radius **r** and height **h**: $V = \frac{1}{3}\pi r^2 h$

- Lateral area of a right circular cone with circumference of the base **c** and slant height **l**: $S = \frac{1}{2}cl$

- Volume of a sphere with radius **r**: $V = \frac{4}{3}\pi r^3$

- Surface area of sphere with radius **r**: $S = 4\pi r^2$

- Volume of a pyramid with base area **B** and height **h**: $V = \frac{1}{3}Bh$

1. What is the perimeter of a right triangle with an angle of 63° and with shorter leg of 11?

(A) 21.6   (B) 21.8   (C) 32.6   (D) 40.1   (E) 56.8

2. If $x + 4y - 6z = 5$ and $2x - 2y + 3z = -10$, then x=?

(A) – 3   (B) – 1   (C) 0   (D) 1   (E) It cannot be determined from the information given.

GO ON TO THE NEXT PAGE ▶▶▶

Model Test 13

3. If a, b, and c are integers and a · b = 2c − 1, then which of the following statements is true?

(A) a is odd and b is even.　　(B) a is even and b is odd.　　(C) a − b is an odd number.

(D) a + b is an even number.　　(E) a and b are both even numbers.

4. The items X, Y and Z are exchangeable in such a way that a of X can be exchanged with b of Y and c of Y can be exchanged with d of Z. If a, b, c, d, and e denote the number of corresponding items then e of X can be exchanged with how many of Z?

(A) $\dfrac{e \cdot d \cdot a}{b \cdot c}$　　　(B) $\dfrac{e \cdot a \cdot c}{d \cdot b}$　　　(C) $\dfrac{e \cdot d \cdot b}{a \cdot c}$　　　(D) $\dfrac{a \cdot c}{e \cdot d \cdot b}$　　　(E) $\dfrac{b \cdot c}{e \cdot d \cdot a}$

5. If sin20° = - sinx then x can be

(A) 70°　　　(B) 160°　　　(C) 220°　　　(D) 250°　　　(E) 340°

6. If $f(a+b)=f(a)+f(b)$ for $a,b \in R$, find f(0) can be

(A) 0　　　(B) 1　　　(C) a　　　(D) b　　　(E) a + b

7 If $\pi^x = \pi + x$, then x can be

(A) -3.11　　　(B) -1.30　　　(C) 0.76　　　(D) 1.31　　　(E) 1.47

GO ON TO THE NEXT PAGE ▶▶▶

Model Test 13

8 If x is defined to be the quantity $\frac{m-n}{mn}$ and n corresponds to one of the numbers defined by the letters A through E as in figure 1, then which of the following can not be correct?

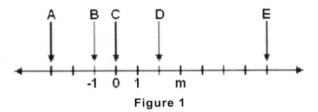

Figure 1

(A) If n = A then x is a negative fraction greater than -1.

(B) If n = B then the absolute value of x is greater than 1.

(C) If n = C then x is an undefined quantity.

(D) If n = D then x is a positive repeating decimal greater than $\frac{2}{13}$.

(E) If n = E then x is a negative number whose decimal part does not have a repeating pattern.

9. For all real numbers b, the equation $\frac{1}{2}x=by-4$ represents which of the following?

(A) A line whose slope is -4.

(B) A line whose y-intercept is -4.

(C) A line whose x-intercept is -8.

(D) A line whose x- intercept is -4.

(E) A vertical line that passes through (4,0).

10. If the shaded region in figure 2 is 84 square units, then a = ?

(A) 2   (B) 3   (C) 4   (D) 6   (E) 7

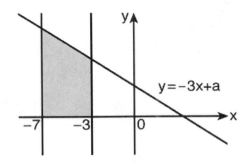

Figure 2

Figure not drawn to scale

11. If $\sec^2x - 3\tan x +1=0$, then tanx can be which of the following?

(A) -3   (B) -1   (C) 1   (D) 3

(E) None of the above

GO ON TO THE NEXT PAGE ▶▶▶

Model Test 13

12. The transverse axis of the hyperbola given in figure 3 is parallel to the x axis. If the equation of line d, which is one of the asymptotes of the hyperbola, is given by the equation y = 0.75x − 4 and point A(4, 2) is one of the vertices then what is the equation of the hyperbola?

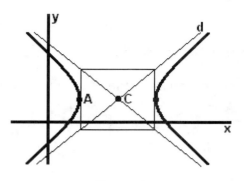

Figure 3
Figure not drawn to scale

(A) $\dfrac{(x-8)^2}{4} - \dfrac{(y-2)^2}{3} = 1$ (B) $\dfrac{(y-8)^2}{16} - \dfrac{(x-2)^2}{9} = 1$

(C) $\dfrac{(x-8)^2}{16} - \dfrac{(y-2)^2}{9} = 1$ (D) $\dfrac{(y-8)^2}{4} - \dfrac{(x-2)^2}{3} = 1$

(E) $\dfrac{(x-8)^2}{16} + \dfrac{(y-2)^2}{9} = 1$

13. What is the amplitude of the function given by $f(x) = -\dfrac{1}{2}\cdot \sin(x)\cdot \cos(x) + 1$?

(A) $\dfrac{1}{4}$ (B) $-\dfrac{1}{2}$ (C) -1   (D) $\dfrac{1}{2}$ (E) 1

14. If cosθ = 0.82, then cos(π - θ)=
(A) -0.57   (B) -0.82   (C) 0.82   (D) 0.57
(E) None of the above

15. For a set of 12 consecutive positive integers, which of the following must be correct?
I. Median is a positive integer.
II. Median equals mean.
III. There is no mode in this set.
(A) I only   (B) II only   (C) III only   (D) II and III only   (E) I, II, and III.

GO ON TO THE NEXT PAGE ▶▶▶

Model Test 13

16. First two terms a_1 and a_2 of a geometric sequence are 5 and 7 respectively. What is the 30th term in this sequence rounded to the nearest integer?

(A) 61,738   (B) 86,433   (C) 86,434   D) 121,007   (E) 121,008

17. If $f(x)$ is defined by $f(x) = (3x)^{(2x)^x}$ then $f(0.8)=$?

(A) 1.61   (B) 3.06   (C) 3.07   (D) 3.57   (E) 3.58

18. If $e^{ix}=\cos(x)+i\sin(x)$ where $i^2 = -1$ then $(-1)^i =$?

(A) $e^{-2\pi}$ (B) $e^{-\pi}$ (C) e^{π} (D) $e^{2\pi}$

(E) None of the above

19. The median of a set of numbers is 5.4. If the smallest number in the set is removed, which of the following could not be the median of the new set?

(A) 4   (B) 5.4   C) 5.6   D) 5.9

(E) There is not enough information given.

20. What is the range of $y = 4 \cdot \cos(3\theta + \pi) - 1$?

(A) $-4 \le y \le 4$ (B) $-3 \le y \le 5$ (C) $-5 \le y \le 3$ (D) $-\dfrac{1}{4} \le y \le \dfrac{1}{4}$ (E) $-\dfrac{5}{4} \le y \le -\dfrac{3}{4}$

GO ON TO THE NEXT PAGE ▶▶▶

Model Test 13

21. Two non parallel lines m and n are in the same coordinate plane and all points on m are reflected across line n. Which point stays fixed?

    I. x – intercept of line m

    II. y – intercept of line n

    III. The point of intersection of m and n

(A) I only   (B) II only   (C) III only   (D) I and II only   (E) I and III only

22. Two points A and B are 10 feet apart in space. How many points in space are 6 feet away from A and 8 feet away from B?

(A) 2   (B) 4   (C) 6   (D) 8   (E) More than 8

23. If $\dfrac{x^2 - 1}{1 - \dfrac{1}{x}} = x^2 + x$ then x can be

(A) any nonzero real number other than 1.   (B) any positive real number

(C) any nonzero real number other than -1.   (D) any nonzero real number.

(E) any real number.

24. What is the domain and range of the function whose graph is given in figure 4?

(A) domain: all reals; range: y < -1 or y > 0

(B) domain: x < -1 or x > 0; range: y ≠ 1

(C) domain: x ∉ {-1, 1}; range: y ∉ [-1,0]

(D) domain: x ≠ 1; range: y < -1 or y > 0

(E) domain: x ≠ 1; range: all reals

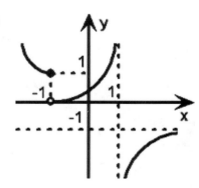

Figure 4

Figure not drawn to scale

GO ON TO THE NEXT PAGE ►►►

25. It is given that the infinite sum $1 - (x+2) + (x+2)^2 - (x+2)^3 + ...$ converges to $\dfrac{1}{x+3}$. The set of all possible values of x are given by

(A) $-3 \le x \le -1$ (B) $-1 < x < 1$ (C) $-3 < x < -1$ (D) $1 < x < 3$ (E) $-3 < x < 1$

26. How many 6 letter words contain distinct letters?

(A) 26^6 (B) $P(26, 6)$ (C) $25 \cdot 26^5$ D) $C(26, 6)$ (E) $26 \cdot 25^5$

27. Given in figure 5 is a right circular cone with a base radius of 1 inch and a slant height of 3 inches. What is the length in inches of the shortest path on the lateral surface of the cone, that joins points A and B?

(A) $\dfrac{\pi}{3}$ (B) $\dfrac{2\pi}{3}$ (C) π (D) $\sqrt{3}$ (E) 3

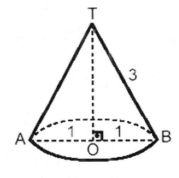

Figure 5

Figure not drawn to scale

28. $\mathbf{A} \cdot \mathbf{X} = \mathbf{B}$ represents the system of equations given where $\mathbf{A}$ is the matrix of coefficients, $\mathbf{X}$ is the matrix of variables and $\mathbf{B}$ is the matrix of results. How many rows and columns does $\mathbf{B}$ have?

$$2x - y = 4$$
$$x + 2y - 3z = 3$$
$$-x + y + z = -3$$

(A) 1 row , 3 columns   (B) 3 rows , 1 column   (C) 3 rows , 3 columns

(D) 3 rows , 5 columns   (E) 5 rows , 3 columns

29. A cube with and edge of length 6 inches is cut into identical smaller cubes each with an edge of length 1 inch as indicated in figure 6. What is the minimum number of cuts needed?

(A) 15   (B) 18   (C) 30   (D) 36   (E) More than 36

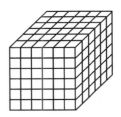

Figure 6

GO ON TO THE NEXT PAGE ▶▶▶

Model Test 13

30. If f(x) satisfies the relation given by $f(a)^n = f(an)$, then f(x) can be

(A) e^x (B) lnx   (C) sinx   (D) tanx   (E) x

31. If the quadratic equation $ax^2+(b+c)x+(d+e)=0$ has two real and distinct roots one of them being zero then which of the following quantities must be nonzero?

(A) a   (B) b   (C) bc   (D) d+e   (E) de

32. If $f(x) = f(x+\pi)$ for all x where f is defined, then f can be?

(A) $f(x) = sinx$ (B) $f(x) = cosx$ (C) $f(x) = tanx$ (D) $f(x)= secx$ (E) $f(x) = cscx$

33. As x approaches 1, $\dfrac{x-1}{\ln x}$ becomes

(A) -1   (B) 0   (C) 1   (D) 2   (E) undefined.

34. The quadratic function f(x) is given by $f(x) = ax^2 + bx + c$ where a, b, and c are all nonzero real numbers. Which of the following gives all possibilities for the number of points that f(x) and |f(x)| can intersect at?

(A) 0 or 1   (B) 0, 1, or more than 2   (C) 0, 1, 2, or more than 2

(D) 0, 1, or 2   (E) 1, 2, or more than 2

35. If both of the points (100, 4) and (0.1, b) are on the graph of $f(x) = \log_a x$ then b = ?

(A) -2   (B) -1   (C) 0   (D) 1   (E) 2

GO ON TO THE NEXT PAGE ▶▶▶

Model Test 13

36. There are totally 24 marbles in a bag of which 7 are blue, 8 are red and 9 are green. If two marbles are selected at random from the bag without replacement, what is the probability that they will have the same color?

(A) $\dfrac{7^2 + 8^2 + 9^2}{48^2}$ (B) $\dfrac{7^2 + 8^2 + 9^2}{24^2}$ (C) $\dfrac{9\cdot 8 + 8\cdot 7 + 7\cdot 6}{48\cdot 47}$ (D) $\dfrac{9\cdot 8 + 8\cdot 7 + 7\cdot 6}{24\cdot 23}$

(E) None of the above.

37. The Cartesian coordinates of a point is (-6, 6). If (r, θ) represents the polar coordinates of this point, (r, θ) can be

(A) $\left(6, \dfrac{3\pi}{4}\right)$ (B) $\left(6\sqrt{3}, \dfrac{3\pi}{4}\right)$ (C) $\left(6\sqrt{2}, \dfrac{-\pi}{4}\right)$ (D) $\left(-6\sqrt{2}, \dfrac{\pi}{4}\right)$ (E) $\left(6\sqrt{2}, \dfrac{-5\pi}{4}\right)$

38. If $x^2 - 3 = x$ then $x + \dfrac{9}{x^2} + 3 = ?$

(A) 7   (B) 14   (C) 21   (D) 23   (E) It cannot be determined from the information given.

39. Graphs of f(x) and g(x) are given. What is the relation between f(x) and g(x)?
(A) g(x)=f(x-1)+1   (B) f(x)+1=g(x+1)
(C) g(x)=f(x+1)+1   (D) f(x)=g(x-1)+1
(E) g(x)+1=f(x+1)

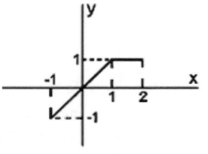

Graph of f(x)

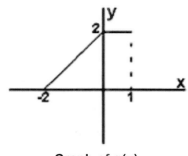

Graph of g(x)

GO ON TO THE NEXT PAGE ▶▶▶

Model Test 13

40. The quadratic function $f(x)$ is given by $f(x) = ax^2 + bx + c$ where a, b, and c are all real numbers. If $f(x)$ and $-f(x)$ intersect at a single point then which of the following must be correct?

(A) c is a positive multiple of $\dfrac{b^2}{4a}$

(B) $c > \dfrac{b^2}{4a}$

(C) $c = \dfrac{b^2}{4a}$

(D) c is a negative multiple of $\dfrac{b^2}{4a}$

(E) $c < \dfrac{b^2}{4a}$

41. A function is said to be even if its graph is symmetric with respect to the y axis. For an even function $f(x)$ which of the following conditions may not always hold?

(A) $f(x)=f(-x)$ (B) $f(x)=f(|x|)$ (C) $f(x)=f(-|x|)$

(D) Its graph is symmetric with respect to the axis defined by the line $y = 0$.

(E) Its graph is symmetric with respect to the axis defined by the line $x = 0$.

42. In figure 7, each number in a cell represents the distance in miles between the cities that are represented by the capital letters in the same row and column as this cell. For example the distance between cities A and D is 130 mi. Assuming that cities A, B, C, D and E are located in a linear path in that order, $x + y = ?$

(A) 90   (B) 100   (C) 120

(D) 130   (E) 140

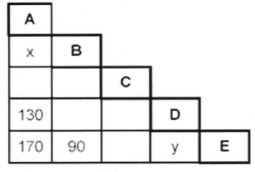

Figure 7

43. A certain brand of wedding ring is made in such a way that 80% of its price is because of the gold and the rest is because of the silver used in producing the ring. The price of gold increases by 12% every year, however the price of silver stays the same. To the nearest dollar, what will be the price of the ring 3 years later if it costs $300 at present?

(A) $300   (B) $324   (C) $397   (D) $421

(E) None of the above

GO ON TO THE NEXT PAGE ▶▶▶

44. If f(x) is defined by the function f(x)=cos(tan(x)), then f(250°)=?

(A) -0.633   (B) -0.923   (C) 0.998   D) 0.999   (E) 1.000

45. For what values of x is the equality given by $\dfrac{\log_5(x^8)}{\log_5(x^6)}=\dfrac{4}{3}$ valid?

(A) For all values of x
(B) For all nonzero values of x
(C) For all values of x except for -1 and 1
(D) For all positive values of x except for 1
(E) For all nonzero values of x except for -1 and 1

46. The area of a certain regular polygon can be calculated by the formula given by $nR^2\sin\left(\dfrac{\pi}{4}\right)$ where R

is the radius of the circle that passes through the vertices of this polygon and n is a positive integer. n=?

(A) 4   (B) 5   (C) 6   (D) 8   (E) 16

47. Which of the following is not correct for the function $f(x) = x^3 + x^2 - 4x - 4$?

(A) It has three real zeros.   (B) All of its zeros are integers.   (C) Its x – intercepts are nonzero.
(D) It is a polynomial function.   (E) Two of its zeros are odd

48. Sevkan wants to use each of the five tires of her car (4 tires and a spare tire) at equal number of miles during a 600 miles trip. How many miles must she use each of the tires?

(A) 120 miles   (B) 150 miles   (C) 240 miles   (D) 480 miles   (E) 600 miles

GO ON TO THE NEXT PAGE ▶▶▶

49. Graph of f(x) is given in figure 8. What will be the graph of -f(-x)?

(A)

(B)

(C)

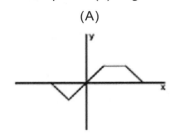
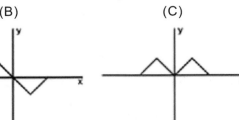
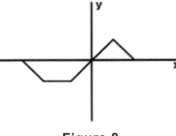

Figure 8

(D)

(E)

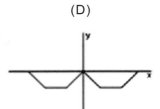
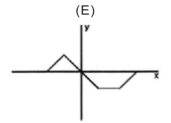

50. Which of the following sets has a greatest element?

I. The set of negative integers

II. The set of negative real numbers

III. Negative multiples of 5

(A) I only   (B) I and II only   (C) II and III only   (D) I and III only   (E) I, II and III

S T O P

END OF TEST

(Answers on page 199 – Solutions on page 237)

Model Test 14

Test Duration: 60 Minutes

Directions: For each of the following problems, decide which is the **best** of the choices given. If the exact numerical value is not one of the choices, select the choice that best approximates this value. Then fill in the corresponding oval on the answer sheet.

Notes:

- A calculator will be necessary for answering some (but not all) of the questions in this test. For each question you will have to decide whether or not you should use a calculator. The calculator you use must be at least a scientific calculator; programmable calculators and calculators that can display graphs are permitted.

- For some questions in this test you may have to decide whether your calculator should be in the radian mode or the degree mode.

- Figures that accompany problems in this test are intended to provide information useful in solving the problems. They are drawn as accurately as possible **except** when it is stated in a specific problem that its figure is not drawn to scale.

- All figures lie in a plane unless otherwise indicated.

- Unless otherwise specified, the domain of any function **f** is assumed to be the set of all real numbers **x** for which **f(x)** is a real number.

Reference Information: The following information is for your reference in answering some of the questions in this test.

- Volume of a right circular cone with radius **r** and height **h**: $V = \frac{1}{3}\pi r^2 h$

- Lateral area of a right circular cone with circumference of the base **c** and slant height **l**: $S = \frac{1}{2}cl$

- Volume of a sphere with radius **r**: $V = \frac{4}{3}\pi r^3$

- Surface area of sphere with radius **r**: $S = 4\pi r^2$

- Volume of a pyramid with base area **B** and height **h**: $V = \frac{1}{3}Bh$

1. If $x = t^5$ and $y = 2t^3 + 1$, what is y in terms of x ?

(A) $x + 1$ (B) $x^2 + 1$ (C) $\sqrt{x} + 1$ (D) $2\sqrt[5]{x^3} + 1$ (E) $\sqrt[5]{x^3} + 1$

2. $\frac{x}{y} + \frac{y}{x} - 1 =$

(A) $\frac{x + y - xy}{xy}$ (B) $\frac{(x-y)^2}{xy}$ (C) $\frac{x + y - 1}{xy}$ (D) $\frac{x^2 + y^2 - 1}{xy}$ (E) $\frac{x^2 + y^2 - xy}{xy}$

GO ON TO THE NEXT PAGE ▶▶▶

Model Test 14

3. If $\dfrac{3}{4}(x^2) = 4$ then what is one possible value of x^3?

(A) .04   (B) 2.31   (C) 5.33   (D) 6.40   (E) 12.32

4. What is the perimeter of a rectangle that has the vertices (-2, -1), (3, -1), and (3, 2)?

(A) 3   (B) 5   (C) 8   (D) 15   (E) 16

5. If $\sin\theta = 0.65$ then $\dfrac{\tan\theta}{\sec\theta} =$

(A) 0.65   (B) 0.76   (C) 0.86   (D) 1.17   (E) 1.33

6. If in figure 2, $\sec\theta = 1.67$ then x = ?

(A) 14.04   (B) 18.78   (C) 29.28

(D) 32.27   (E) 39.16

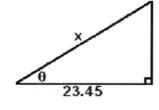

Figure 1

7. If $f(x) = x^2+2$ and $g(x) = \sqrt[3]{x} + 1$, what is g(f(5)) + f(g(8))?

(A) 3   (B) 4   (C) 11   (D) 15   (E) 27

GO ON TO THE NEXT PAGE ▶▶▶

8. What is the equation of the tangent to the curve given by $(x - 3)^2 + (y + 2)^2 = 25$ at the point whose x coordinate is 6 if this point lies on this curve in the first quadrant?

(A)   (B)   (C)   (D)   (E)

$y + 2 = \dfrac{4}{3}(x + 6)$ $y + 2 = \dfrac{4}{3}(x + 6)$ $y = -x + 8$ $y + 2 = -\dfrac{3}{4}(x - 3)$ $y - 2 = -\dfrac{3}{4}(x - 6)$

9. Line l is given by $y = -3x + b$. If line m is perpendicular to line l, then line m can be

(A) $y = 3x + b$ (B) $y = \dfrac{1}{3}x + b$ (C) $y = -3x - b$ (D) $y = 3x - b$ (E) $y = -\dfrac{1}{3}x + b$

10. One period of a periodic function f(x) is shown in figure 2.
What is the range of f(x)?

(A) $y = 0$

(B) $0 \le x \le 2\pi$

(C) $0 \le x \le 4\pi$

(D) $-2 \le y \le 2$

(E) All real numbers

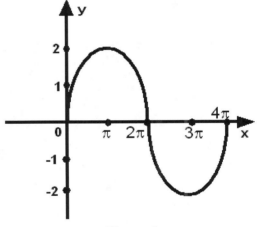

Figure 2
Figure not drawn to scale

11. When it's defined, $\dfrac{1 + \cot^2 \theta}{\cot^2 \theta} =$

(A) 0   (B) 1   (C) $\sec^2 \theta$ (D) $\cos^2 \theta$ (E) $1 - \cos^2 \theta$

GO ON TO THE NEXT PAGE ▶▶▶

12. The graph given in figure 3 can be represented by the equation

(A) $x^2 + y^2 = 1$.

(B) $y = \sqrt{1-x^2}$.

(C) $x = \sqrt{1-y^2}$.

(D) $y = -\sqrt{1-x^2}$.

(E) $x = -\sqrt{1-y^2}$.

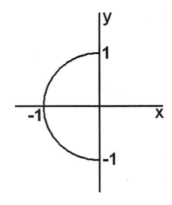

Figure 3

13. Which of the following functions has exactly two zeros at -3 and 2?

(A) $f(x) = \dfrac{(x-1)(x-3)(x+2)}{(x+4)(x+5)}$

(B) $f(x) = \dfrac{(x-1)(x+4)(x+5)}{(x+3)(x-2)}$

(C) $f(x) = \dfrac{(x-1)(x+3)(x-2)}{(x+5)(1-x)}$

(D) $f(x) = \dfrac{(x-1)(x+3)(x-2)}{(x+3)(x+5)}$

(E) $f(x) = \dfrac{(x-1)(x+1)(x+2)}{(x+3)(x-2)}$

14. A cubic function of x is given by $f(x) = ax^3 + bx^2 + cx + d$ where a, b, c, and d are real numbers and $a \neq 0$. If the function given by $f(x) - k$ has a single zero for all real values of k, then which of the following cannot be correct?

(A) f(x) is an invertible function.

(B) f(x) is a one to one function.

(C) f(x) is a decreasing function.

(D) f(x) intersects the y axis exactly once.

(E) f(x) intersects the x axis more than once.

15. If $f(x) = x - 2$, which of the following equals $\dfrac{f(8) \cdot (f(5) - f(4))}{f(4)}$?

(A) f(1)

(B) f(3)

(C) f(5)

(D) f(15)

(E) f(5) − f(3)

GO ON TO THE NEXT PAGE ▶▶▶

Model Test 14

16. A sphere and a cylinder have equal volumes and radii. If the radius of each solid is 5, what is the total area of the cylinder?

(A) 157.08   (B) 209.44   (C) 287.98   (D) 350.00   (E) 366.52

17. In figure 4, if θ and β measure 40° and 85° respectively then what is the length of AC?

(A) 5.22   (B) 5.23   (C) 5.82

(D) 6.66   (E) 6.67

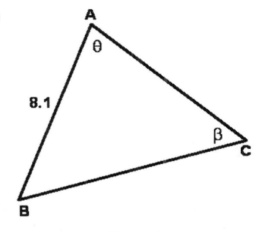

Figure 4

Figure not drawn to scale

18. The median of a set of numbers is 3.8. If the smallest number in the set is removed, which of the following could not be the median of the new set?

(A) 3   (B) 3.8   (C) 4.2   (D) 5.4   (E) There is not enough information given.

19. If $x^2 - 2x - 3 < 0$ and $f(x) = -x^2 + 2x + 1$, then

(A) $f(x) < -1$ or $f(x) > 3$ (B) $-2 < f(x) < 0$ (C) $-2 < f(x) < 2$

(D) $-2 < f(x) \leq 2$ (E) $f(x) > 0$

20. If $-3 \leq x \leq 1$, and $f(x) = \left| x^2 - 4 \right|$, then what is the range of f(x)?

(A) $y \geq 0$ (B) $0 \leq y \leq 9$ (C) $0 \leq y \leq 5$ (D) $4 \leq y \leq 9$ (E) $y \leq 9$

GO ON TO THE NEXT PAGE ▶▶▶

21. A rat will go from A to B along the grid lines in figure 5 and it is allowed to move 1 unit up or one unit right in each move. How many different paths are there for the rat?

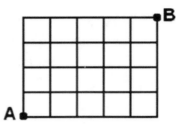

Figure 5

(A) P(9,4) (B) 4^5 (C) 5·4 (D) C(9,4) (E) 5^4

22. The third and the sixth terms of a geometric sequence are 4 and -108 respectively. What is the second term in this sequence?

(A) – 12 (B) – 4/3 (C) – 3/4 (D) 4/9 (E) 36

23. For which of the following values of x is cotx always greater than sinx?

(A) 0° < x < 45° (B) 45° < x < 90° (C) 90° < x < 135° (D) 135° < x < 180°
(E) Not possible

24. Which of the following functions does not have an inverse?

(A) $f(x) = \dfrac{1}{3x}$ (B) $f(x) = (3x)^2$ (C) $f(x) = (3x)^3$ (D) $f(x) = e^{-x}$ (E) $f(x) = \ln(3x)$

25. What is the domain of the function $f(x) = \dfrac{1}{x^3 - 4x^2 + 4x}$?

(A) x < 0 (B) x ≠ 0, 2 (C) 0 < x < 2 (D) 0 ≤ x ≤ 2 (E) All real numbers

GO ON TO THE NEXT PAGE ▶▶▶

Model Test 14

26. For all θ, $(e^{\sin(-\theta)}) \cdot (e^{-\sin(\theta)}) =$

(A) 1   (B) e   (C) $e^{\sin 2\theta}$ (D) $e^{-2\sin\theta}$ (E) $e^{\sin^2\theta}$

27. The graph of $y = f(x)$ is shown in figure 6. Which of the following statements must be true?

I. $|f(x)| > 0$ II. $f(x) = -f(-x)$ III. If $x_1 \le x_2$ then $f(x_1) \le f(x_2)$

(A) I only   (B) II only   (C) III only   (D) I and II only   (E) II and III only

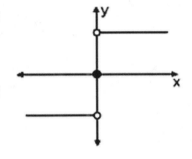

Figure 6

28. What are all values of x for which $(x^2+1)(2x-1) + (x - x^2)(2x)$ equals 0?

(A) $x = 1 \pm \sqrt{2}$ (B) $x = -1 \pm \sqrt{2}$ (C) $x = 1 \pm i\dfrac{\sqrt{2}}{2}$ (D) $x = 1; \ 1 \pm i\sqrt{2}$

(E) None of the above

29. The function given by $f(x) = \dfrac{2x+4}{x-r}$ has no vertical asymptotes. What is the value of r?

(A) -2   (B) -1   (C) 0   (D) 1   (E) 2

30. If $f(x) = -x^3 + x^2 + 6x$, for what values of x is $f(x) \le 0$?

(A) $-2 \le x \le 0$ or $x \ge 3$ (B) $-1 \le x \le 1$ (C) $2 \le x \le 4$ (D) $x \le -2$ or $0 \le x \le 3$ (E) $0 \le x \le 1$

GO ON TO THE NEXT PAGE ▶ ▶ ▶

31. What is the equation for the set of points at a distance of 1 unit from the origin?

(A) $x^2 + y^2 - 1 = 0$ (B) $(x - 1)^2 + (y - 1)^2 = 0$ (C) $(x - 1)^2 + (y + 1)^2 = 0$

(D) $x^2 + y^2 + 1 = 0$ (E) $(x + 1)^2 + (y - 1)^2 = 1$

32. What is the range of $f(x) = \begin{cases} x + 3 & x \ge 0 \\ -x^2 & x < 0 \end{cases}$?

(A) $y < 0$ (B) $0 \le y \le 3$ (C) $y \ge 3$ (D) $y < 0$ or $y \ge 3$ (E) All real numbers

33. If $f^{-1}(x) = 0.5x^2$ for $x > 0$, then $f(3) =$

(A) 0.45   (B) 1.15   (C) 1.75   (D) 2.45   (E) 4.5

34. For what values of x is the inverse of $f(x) = \dfrac{1}{x - 1}$ not defined?

(A) x = 0 only   (B) x = 1 only   (C) x < 1   (D) x = 0 or x = 1   (E) x > 1

35. If the terms of an arithmetic sequence are denoted as a_n where $a_7 = 4$ and $a_{10} = 13$, then what is a_{72}?

(A) 178   (B) 196   (C) 199   (D) 202   (E) 220

GO ON TO THE NEXT PAGE ▶▶▶

Model Test 14

36. The set of points (x, y) in the coordinate plane that satisfy the relation given by $\left(\sqrt{x^2 + y^2} - 1\right)(x + y - 1) \le 0$ are correctly described by which of the following?

(A)   (B)   (C)   (D)   (E)

37. Vectors **x** and **y** have the components of (6, 8) and (-1, -16) respectively. What is the magnitude of the vector **z** defined by **z = y − x**?

(A) 9.43   (B) 10.00   (C) 18.03   (D) 25.00

(E) None of the above

38. $\cos^{-1}(\sin 150°) =$

(A) 0°   (B) 0.5   (C) 30°   (D) 60°   (E) 90°

39. Which of the following operations does not change the standard deviation in an array of numbers?

   I.   Add 1 to each number

   II.   Subtract 2 from each number

   III. Multiply each number by 1

   IV. Divide each number by 3

(A) I or III only   (B) II or III only   (C) I or II only   (D) III or IV only   (E) I, II or III only

GO ON TO THE NEXT PAGE ►►►

Model Test 14

40. A right triangle in 3-dimensional space is formed by the points (3, 2, 4), (3, 2, -1) and (0, 2, -1). The length of the hypotenuse is

(A) 3.00   (B) 5.00   (C) 5.83   (D) 6.40   (E) 7.51

41. Which of the following expresses the polar point $\left(4, \dfrac{5\pi}{6}\right)$ in rectangular coordinates?

(A) (-2, 3.46)   (B) (-3.46, 2)   (C) (3.46, -2)   (D) (2, -3.46)   (E) (-4, 2.62)

42. $\dfrac{(2n+4)!}{(2n+2)!} =$

(A) 2   (B) n + 1   (C) n + 2   (D) $\dfrac{n+2}{n+1}$ (E) (2n + 4)(2n + 3)

43. A circle with equation $(x + 2)^2 + (y - 2)^2 = 25$ is rotated around the line y = - x. What is the volume of the solid that is created?

(A) 104.72   (B) 150.80   (C) 268.03   (D) 523.60   (E) 1570.80

44. If $f(x,y) = \begin{cases} x^3 - 1 & x \le y \\ x + y & x > y \end{cases}$, what is the value of $f(-1,1) - f(1,-1)$?

(A) -4   (B) -2   (C) 0   (D) 2   (E) 4

GO ON TO THE NEXT PAGE ►►►

Model Test 14

45. If the graph of $f(x) = 3(x + 1)^2 - 5$ is translated 5 units up and 1 units right, what is the y coordinate of the point on the transformed graph whose x coordinate is 1?

(A) 3   (B) 5.10   (C) 6   (D) 17   (E) 27

46. The height of a cylinder is five times its radius and the lateral area of the cylinder is 3000 square inches. What is the volume of the cylinder in cubic feet?

(A) 5.8   (B) 8.5   (C) 9.7   (D) 8.4   (E) 7.9

47. An indirect proof of the statement "if a is a member of set A, then a is not a member of set B" could begin with the assumption that

(A) a is a member of set A   (B) a is a member of set B   (C) a is not a member of set A

(D) a is not a member of set B   E) a is neither a member of A or B

48. A relation β is defined in the set of lines l_1 through l_7 given in figure 7 above as follows: $\beta = \{(l_n, l_{n+1}): l_n \parallel l_{n+1}\}$. How many elements does β have?

(A) 2   (B) 3   (C) 4

(D) 5   (E) 6

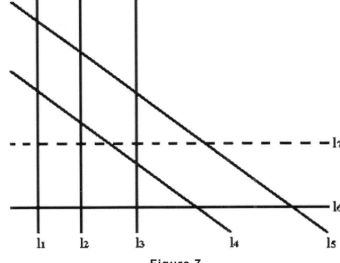

Figure 7

GO ON TO THE NEXT PAGE ▶▶▶

Model Test 14

49. A function f(x) is said to be even if f(-x) = f(x) for all x in the domain. Which of the following functions is even?

(A) $f(x) = \dfrac{1}{x}$ (B) $f(x) = \sin x$ (C) $f(x) = |3x|$ (D) $f(x) = x(x+1)$ (E) $f(x) = x$

50. In figure 8, circle C is tangent to side AB of the isosceles triangle ABC. If $\angle ACB = 120°$ and $AB = 12\sqrt{3}$ then what is the area of the shaded region?

(A) 6.10   (B) 12.33   (C) 18.20

(D) 20.14   (E) 24.65

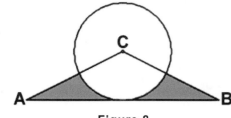

Figure 8
Figure not drawn to scale

S T O P

END OF TEST

(Answers on page 199 – Solutions on page 240)

Model Test 15

Test Duration: 60 Minutes

Directions: For each of the following problems, decide which is the **best** of the choices given. If the exact numerical value is not one of the choices, select the choice that best approximates this value. Then fill in the corresponding oval on the answer sheet.

Notes:

- A calculator will be necessary for answering some (but not all) of the questions in this test. For each question you will have to decide whether or not you should use a calculator. The calculator you use must be at least a scientific calculator; programmable calculators and calculators that can display graphs are permitted.

- For some questions in this test you may have to decide whether your calculator should be in the radian mode or the degree mode.

- Figures that accompany problems in this test are intended to provide information useful in solving the problems. They are drawn as accurately as possible **except** when it is stated in a specific problem that its figure is not drawn to scale.

- All figures lie in a plane unless otherwise indicated.

- Unless otherwise specified, the domain of any function **f** is assumed to be the set of all real numbers **x** for which **f(x)** is a real number.

Reference Information: The following information is for your reference in answering some of the questions in this test.

- Volume of a right circular cone with radius **r** and height **h**: $V = \frac{1}{3}\pi r^2 h$

- Lateral area of a right circular cone with circumference of the base **c** and slant height **l**: $S = \frac{1}{2}cl$

- Volume of a sphere with radius **r**: $V = \frac{4}{3}\pi r^3$

- Surface area of sphere with radius **r**: $S = 4\pi r^2$

- Volume of a pyramid with base area **B** and height **h**: $V = \frac{1}{3}Bh$

1. If a and b are distinct prime numbers, which of the following numbers must be odd?

(A) ab   (B) 4a+b   (C) a+b+5   (D) ab-1   (E) 2(a+1)-2b+1

2. If $3x^5 = 4$, then $5(3x^5)^2 =$

(A) 40   (B) 60   (C) 80   (D) 100

(E) None of the above

GO ON TO THE NEXT PAGE ▶▶▶

Model Test 15

3. The slopes of two lines that do not intersect can be

   I. equal

   II. reciprocals of each other

   III. negative reciprocals of each other

(A) I only   (B) I or II only   (C) I or III only   (D) II or III only   (E) I, II or III

4. If $x^2 + y^2 = 5$ and $x^2 - y^2 = 3$, then x can be

(A) -1   (B) -2   (C) 3   (D) 4   (E) 5

5. If $\frac{5}{6}x \neq 0$, then $\frac{5}{6} - x$ cannot be

(A) 0   (B) $\frac{5}{6}$ (C) $\frac{6}{5}$ (D) $\frac{25}{36}$ (E) 1

6. $x\left(\frac{2}{y} - \frac{2}{z}\right) = ?$

(A) $\frac{2xz - 2xy}{yz}$ (B) $\frac{2x}{y - z}$ (C) $\frac{x}{yz}$ (D) $\frac{x}{y - z}$ (E) $\frac{2}{xy - xz}$

7. If $2^{x-1} = A$ and $A = \frac{1}{3}$; x = ?

(A) -1.58   (B) -0.58   (C) -0.85   (D) 0.58   (E) 0.85

GO ON TO THE NEXT PAGE ▶▶▶

Model Test 15

8. log(sin20)+ log(sin20°)=?

(A) -0.465　　　　(B) -0.466　　　　(C) -0.505　　　　(D) -0.506　　　　(E) -2.681

9. If all variables in the equations xy=35 and yz=45, represent integers less than 1, then what is the maximum value of x + y + z?

(A) -9　　　　(B) -21　　　　(C) -35　　　　(D) -45　　　　(E) -81

10. If $(- 4x)^{2k-1} < 0$, where x is a real number and k is a negative integer, then

(A) x < 0　　　(B) x ≤ 0　　　(C) x > 0　　　(D) x ≥ 0　　　(E) x is any real number

11. Given that $A@B = A^{B} - B^{A}$ and 2@C = 0 then C cannot be

(A) irrational.　　　(B) negative.　　　(C) positive.　　　(D) an integer.　　　(E) zero.

12. Two quantities A and B are related by the linear regression model given by the following equation:
A = - 3.05B + 7.35. Which of the following can be deduced?

　　I. There is a positive correlation between A and B.

　　II. When B is less than 20% the predicted minimum integer value of A will be 7.

　　III. The slope indicates that as B is increased by 1, A is decreased by 3.05.

(A) II only　　　(B) I and II only　　　(C) II and III only　　　(D) I and III only　　　(E) I, II, and III

GO ON TO THE NEXT PAGE ▶▶▶

Model Test 15

13. Given that $z = t^3 - 1$ and $y = 3t^2 + 4$. If $y - z = 9$ then t can be?

(A) -1   (B) 0   (C) 1   (D) 2   (E) 3

14. Which of the following relations can represent a function?

    I. $\beta = \{(x,y) | x^2 = y^3;\ x,y \in R\}$

    II. $\beta = \{(x,y) | x^3 = y^2;\ x,y \in R\}$

    III. $\beta = \{(x,y) | x^2 + y^2 = 0;\ x,y \in R\}$

(A) I only   (B) II only   (C) III only   (D) I and III only   (E) I, II and III

15. If in figure 1 BC is given to be x then which of the following is equal to AC?

(A) $x \cdot \sin(\theta)$ (B) $x \cdot \sin(\pi - \theta)$ (C) $\dfrac{x}{\cos(\theta)}$

(D) $\dfrac{x}{\sin(\pi - \theta)}$ (E) $\dfrac{x}{\sin(\theta - \pi)}$

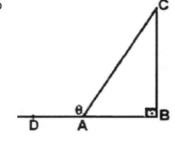

Figure 1

16. If $f(x) = \sqrt{3x - 8}$ and $g(x) = x^3 + x + 1$, then $f(g(2)) = $?

(A) ±5   (B) 5   (C) 1/5   (D) ±1/5

(E) none of the above

17. Which of the following is the set of all possible x values for the function given by $f(x) = \log_{(5-x)}(x - 1)$?

(A) x > 1 or x < 5   (B) 0 < x < 5 and x ≠ 4   (C) 1 < x < 5 or x ≠ 4

(D) 1 < x < 5 and x ≠ 4   (E) 1 ≤ x ≤ 5 and x ≠ 4

GO ON TO THE NEXT PAGE ▶▶▶

Model Test 15

18. Which of the following is the inverse of the function given in the table?

x	f(x)
1	2
2	3
3	4
4	1

(A)

x	$f^{-1}(x)$
1	2
2	3
3	4
4	1

(B)

x	$f^{-1}(x)$
1	4
2	1
3	2
4	3

(C)

x	$f^{-1}(x)$
1	1
2	2
3	3
4	4

(D)

x	$f^{-1}(x)$
1	2
2	3
3	1
4	4

(E)

x	$f^{-1}(x)$
1	4
2	3
3	2
4	1

19.   Statement 1: No Zandies are Mandies;

   Statement 2: All Mandies are Tandies.

Which of the following must be correct?

(A) Some Tandies are not Zandies.   (B) Some Mandies are Zandies.

(C) Some Tandies are Zandies.   (D) All Tandies are Mandies.

(E) A Tandy that is not a Mandy cannot be a Zandy.

20. sinx = -sinα and cosx = cosα are given. What is x in terms of α?

(A) $2\pi + \alpha$ (B) $2\pi - \alpha$ (C) $\alpha - \pi/2$ (D) $\alpha + \pi/2$ (E) $\pi/2 - \alpha$

21. If $f(x) = x^2 - 9$ and $f(f(A)) = 0$ then A cannot be

(A) -3.5   (B) -2.4   (C) 2.3   (D) 2.4   (E) 3.5

GO ON TO THE NEXT PAGE ▶▶▶

Model Test 15

22. Statement: The cube of a number x is less than itself.

The set of all x values that satisfy the above statement are included in the set that contains

(A) the negative real numbers greater than -1.   (B) the real numbers less than -1.

(C) the positive real numbers less than 1.   (D) the real numbers less than 1.

(E) the real numbers greater than 1.

23. If g(x) is given in terms of an arbitrarily selected function f(x) and g(x) = g(-x) then g(x) can be defined for all f(x) as

    I. $g(x) = f^2(x)$ II. $g(x) = f(x) + f(-x)$ III. $g(x) = f(-|x|)$

(A) I only   (B) II only   (C) I and II only   (D) II and III only   (E) I, II and III

24. Point A has the coordinates (3, 15) and point B is on the circle that has an equation given by $(x - 8)^2 + (y - 3)^2 = 4$. What is the minimum distance between A and B?

(A) 9   (B) 11   (C) 13   (D) 15   (E) It cannot be determined from the information given.

25. $f(x,y) = \sqrt{3x^2 - 4y}$ and $g(x) = 3^x \Rightarrow g(f(2,1)) = ?$

(A) 22.3   (B) 22.4   (C) 12.72   (D) 12.73   (E) 6561

26. A parabola given by $f(x) = ax^2 + bx + c$ passes through the points (1, 4), (0, 3), and (-1, 6). Which of the following is a correct statement for this parabola?

(A) It opens downward.   (B) It is tangent to the x axis.

(C) It is a positive function of x.   (D) It has two real and unequal zeros.

(E) Its axis of symmetry is x = -0.25.

GO ON TO THE NEXT PAGE ▶▶▶

27. What is the equation of the locus of points at a distance of 10 from the point (3,-1)?

(A) $(x-3)^2+(y+1)^2=10$ (B) $(x+3)^2+(y-1)^2=10$ (C) $(x-3)^2+(y+1)^2=100$

(D) $(x+3)^2+(y-1)^2=100$ (E) $(x+1)^2+(y-3)^2=100$

28. Base of the right pyramid given in figure 2 is a regular hexagon. If AB=6 inches and BC=4 inches, then what is the volume of the pyramid in cubic inches?

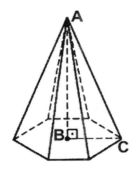

(A) $12\sqrt{3}$ (B) $24\sqrt{3}$ (C) $48\sqrt{3}$

(D) $96\sqrt{3}$ (E) $192\sqrt{3}$

Figure 2

Figure not drawn to scale

29. The first three terms of an arithmetic sequence are 4t, 10t-1, and 12t+2. What is the numerical value of the fiftieth term?

(A) 74   (B) 79   (C) 244   (D) 249   (E) 254

30. For a polynomial function P(x) what is the remainder when P(x-2) is divided by x+1?

(A) P(-3)   (B) P(-1)   (C) P(3)   (D) P(1)

(E) None of the above.

31. For the trapezoid given in figure 3, AD=4, DC=6 and ∠ CBA measures 45°. If the trapezoid is rotated 180° about AB, then what will be the volume of the resulting solid?

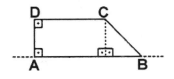

(A) 368   (B) 502   (C) 369

(D) 503   (E) 184

Figure 3

Figure not drawn to scale

GO ON TO THE NEXT PAGE ▶▶▶

Model Test 15

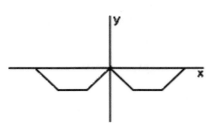

32. Graph of f(|x|) is given above. Which of the following can the graph of f(x)?

(A) 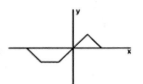 (B) 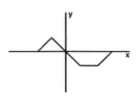 (C) 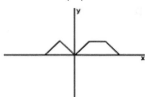 (D) 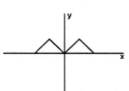 (E)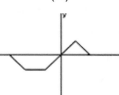

33. What is the largest possible domain for the function $f(x) = \sqrt{x+4} + \dfrac{1}{x-3}$?

(A) x > 4 and x ≠ 3   (B) x ≥ - 4 or x ≠ 3   (C) x > - 4 and x ≠ 3

(D) x ≥ - 4 or x ≠ 3   (E) x ≥ - 4 and x ≠ 3

34. The two concentric circles given in figure 4 have radii of 5 and 3 inches and their centre is at O. If measure of angle ∠AOD is $\dfrac{2\pi}{5}$ radians, then what is the area of the shaded region in square inches?

(A) $\dfrac{4\pi}{5}$ (B) $\dfrac{8\pi}{5}$ (C) $\dfrac{16\pi}{5}$ (D) $\dfrac{12\pi}{5}$ (E) 4π

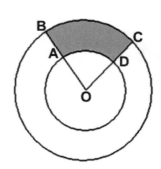

Figure 4
Figure not drawn to scale

35. The point A(3, 5) and B(m, n) are symmetric in the line 3x − 4y + 1 = 0. What is the length of segment AB?

(A) 2   (B) 4   (C) 6   (D) 8   (E) 10

GO ON TO THE NEXT PAGE ▶▶▶

Model Test 15

36. What is the obtuse angle between the lines y = x + 3 and y = -$\sqrt{3}$ x + 2?

(A) 45°   (B) 75°   (C) 105°   (D) 120°   (E) 135°

37. For the function defined by $f(x) = A \cdot \cos(Bx + C) + D$, coefficients A, B, C and D are all real numbers greater than 2. Increasing A would effect which of the following attributes of f(x)?

(A) Amplitude only

(B) Vertical shift only

(C) Period and frequency only

(D) Horizontal (phase) shift only

(E) Period, frequency and phase shift only

38. In a class, there are 23 boys and 19 girls. 10 of the boys and 7 of the girls have participated in the talent show. If a student is selected at random what is the probability that the student has participated in the talent show or he is a boy?

(A) 5/7   (B) 10/21   (C) 10/17   (D) 5/21   (E) 10/23

For questions 39 and 40, please refer to the periodic function given in figure 5. Please note that the frequency of this function is $\frac{1}{6}$.

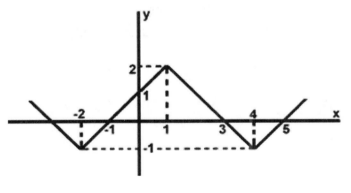

Figure 5

39. What is the amplitude of the function given in figure 5?

  (A) 1/6   (B) 6.0   (C) 3.0   (D) 1.5   (E) 0.5

40. What is the x-coordinate of the least 10th positive zero of the function given in figure 5?

(A) 24   (B) 26   (C) 27   (D) 29   (E) 33

GO ON TO THE NEXT PAGE ▶▶▶

41. At a certain harbor, the depth d feet of the water level at time t hours after midnight is modeled by

$d(t) = A + B \cdot \sin\left(\dfrac{\pi}{2} - \dfrac{\pi t}{6}\right)$, $0 \le t \le 24$, where A, and B are positive constants. P(6, 30) and Q(12, 50) are

minimum and maximum points on the graph of d(t) respectively. What is the first time in the 24 hour period when the depth of water is 35 ft?

(A) 4 AM   (B) 8 AM   (C) 4 PM   (D) 8 PM

(E) None of the above

42. $f(x)=2x^2+12x+3$. If the graph of f(x-k) is symmetric about the y axis, what is k?

(A) -3   (B) 3   (C) -15   (D) 15

(E) None of the above

43. In figure 6, f(x) is a continuous function and g(x) is a piecewise function with one point of discontinuity. Which of the following is false?

(A) (f+g)(x) has at least one zero.

(B) (f-g)(x) is always nonnegative.

(C) (f.g)(x) has one point of discontinuity.

(D) $\left(\dfrac{f}{g}\right)$(x) has one point of discontinuity.

(E) The equation f(x) = g(x) has exactly one root.

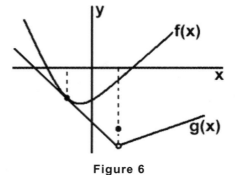

Figure 6

44. What is the real polynomial of the lowest degree whose roots include -2 and 3i where $i=\sqrt{-1}$?

(A) $x^3-2x^2-9x-18$ (B) $x^3-2x^2+9x+18$ (C) $x^3+2x^2-9x+18$

(D) $x^3-2x^2+9x-18$ (E) $x^3+2x^2+9x+18$

GO ON TO THE NEXT PAGE ▶▶▶

Model Test 15

45. If 5ED = 4AE = 2AB = AC = 300 feet in figure 7, then what is the area in square feet of the shaded triangular field?

(A) 13.500   (B) 13,500   (C) 27,000   (D) 36,000

(E) It cannot be determined from the given information

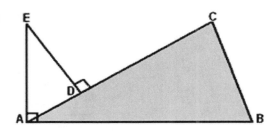

Figure 7

Figure not drawn to scale

46. Given that f is odd and g is even and that f(a)=b=g(c) then what is the value of $\dfrac{f(-a)}{g(-c)}+f(-a)-g(-c)$?

(A) 2b − 1   (B) -2b + 1   (C) -2b − 1   (D) -1   (E) 1

47. Given two sets of data A={1.0, 2.0, 3.0, 4.0, 5.0} and B={2.0, 2.5, 3.0, 3.5, 4.0}. Which of the following is false?

(A) The range of data set A is greater than that of B.

(B) Both sets of data have the same mean and median.

(C) Each data set contains terms of an arithmetic sequence.

(D) Standard deviation of the data set A is less than that of B.

(E) A∩B has 8 subsets.

48. If f(x,y)=x^2-xy+y^2 for all real numbers x and y, which of the following are correct?

I. f(x,y)=f(x,-y)

II. f(x,y)=f(-x,y)

III. f(x,y)=f(-x,-y)

(A) I only   (B) II only   (C) III only   (D) I and III only   (E) I, II and III

GO ON TO THE NEXT PAGE ▶▶▶

Model Test 15

The sales of cars at the Gofast Car Factory

	Month 1	Month 2	Month 3
Model A	25	32	17
Model B	21	23	19
Model C	19	11	10

49. The entries in the table above represent the number of cars that were sold during a three-month sale at the Gofast Car Factory. The prices of models A, B, and C are $59k, $79k, and $99k, respectively, where k denotes thousands of US dollars. Which of the following matrix representations, gives the total income, in k thousands of dollars received from the sale of the cars for each of the three months?

(A) $\begin{bmatrix} 25 & 32 & 17 \\ 21 & 23 & 19 \\ 19 & 11 & 10 \end{bmatrix}\begin{bmatrix} 59 & 79 & 99 \end{bmatrix}$ (B) $\begin{bmatrix} 25 & 32 & 17 \\ 21 & 23 & 19 \\ 19 & 11 & 10 \end{bmatrix}\begin{bmatrix} 59 \\ 79 \\ 99 \end{bmatrix}$ (C) $\begin{bmatrix} 59 & 79 & 99 \end{bmatrix}\begin{bmatrix} 25 & 32 & 17 \\ 21 & 23 & 19 \\ 19 & 11 & 10 \end{bmatrix}$ (D) $\begin{bmatrix} 59 \\ 79 \\ 99 \end{bmatrix}\begin{bmatrix} 25 & 32 & 17 \\ 21 & 23 & 19 \\ 19 & 11 & 10 \end{bmatrix}$

(E) None of the above

50. The set of points (x,y) given by $x^2-4y^2-4x-24y-32=0$ represents

(A) a hyperbola.   (B) an ellipse.   (C) a parabola.   (D) a circle.   (E) two lines.

STOP

END OF TEST

(Answers on page 199 – Solutions on page 243)

S T O P

ANSWERS TO MODEL TESTS

Model Test 1 Answers

#		#		#		#		#	
1	B	*11	E	21	C	31	A	41	D
2	A	12	D	22	D	32	A	42	B
3	A	*13	C	23	B	33	D	43	D
4	E	14	B	24	B	34	C	44	E
5	D	15	E	25	E	35	B	45	B
6	B	16	B	26	A	*36	D	46	C
7	B	17	B	27	E	37	B	47	C
8	C	*18	D	28	E	38	C	48	E
9	D	19	D	29	C	39	E	*49	D
10	D	20	A	30	B	40	E	50	D

Model Test 2 Answers

#		#		#		#		#	
1	A	11	C	21	B	31	B	41	C
2	C	12	B	22	D	32	E	42	B
3	C	13	E	23	B	33	E	*43	B
4	D	14	D	*24	C	34	B	44	A
*5	A	15	A	*25	C	35	E	45	E
*6	C	16	A	26	A	*36	B	*46	D
7	A	*17	B	27	E	37	A	47	D
8	C	*18	C	28	D	38	D	*48	C
9	A	19	C	29	B	39	C	49	B
10	E	20	E	30	B	*40	C	50	E

Model Test 3 Answers

#		#		#		#		#	
1	B	11	B	*21	D	31	B	41	E
2	E	12	D	22	D	*32	D	*42	C
3	A	13	D	23	C	33	D	*43	C
4	C	14	E	24	E	34	A	44	C
5	C	15	E	25	E	*35	B	45	A
6	D	16	B	26	E	36	C	46	C
7	D	17	E	27	C	37	D	47	C
8	A	18	A	28	E	38	E	48	A
9	D	19	D	29	B	39	C	49	A
10	A	*20	B	30	B	40	C	*50	D

Model Test 4 Answers

#		#		#		#		#	
1	A	11	D	21	C	31	D	*41	E
2	D	12	C	22	C	32	D	42	C
3	A	13	A	23	D	33	D	*43	E
4	E	14	B	24	E	34	E	44	D
5	D	15	C	25	E	35	B	45	D
6	E	*16	A	*26	C	*36	E	46	B
7	D	17	D	27	C	37	C	*47	D
8	C	18	D	28	E	*38	E	48	E
9	B	19	D	29	D	39	E	*49	D
10	C	20	B	*30	C	40	A	50	C

Model Test 5 Answers

#		#		#		#		#	
1	C	11	D	21	A	31	B	*41	B
2	C	12	E	22	B	*32	D	*42	C
3	C	13	D	*23	A	33	E	*43	B
4	E	*14	B	24	E	34	D	44	B
5	B	15	D	25	D	*35	C	*45	B
6	B	16	A	26	D	*36	D	46	A
7	B	17	E	27	B	37	D	47	D
*8	B	18	C	28	C	38	D	48	B
*9	C	*19	E	29	E	39	A	49	D
10	E	20	C	30	D	*40	E	50	C

Model Test 6 Answers

#		#		#		#		#	
1	D	11	C	*21	C	31	C	*41	B
2	D	12	D	22	E	32	A	42	E
*3	A	13	C	23	A	33	B	43	C
4	B	14	C	24	C	*34	C	44	D
5	D	15	A	25	B	35	D	*45	B
6	C	16	D	26	B	*36	B	46	D
*7	E	17	C	27	D	37	E	47	B
8	A	18	D	28	A	38	D	*48	D
9	E	19	B	29	A	39	B	49	C
10	B	*20	D	30	C	40	D	50	C

Model Test 7 Answers

#		#		#		#		#	
1	B	11	D	21	C	*31	D	41	A
2	E	12	B	22	A	32	B	42	C
3	C	13	B	23	C	33	D	43	D
4	A	14	B	24	D	*34	C	44	E
*5	E	15	D	25	E	35	E	45	C
6	E	16	D	26	C	36	D	46	E
7	D	17	C	27	C	37	A	*47	A
8	C	*18	B	28	B	38	C	48	D
9	E	19	E	29	C	39	C	49	C
10	C	20	C	30	C	*40	D	50	C

Model Test 8 Answers

#		#		#		#		#	
1	E	11	D	21	C	31	D	41	A
2	D	12	D	22	C	32	B	42	B
3	B	13	A	23	A	33	E	43	A
4	E	14	B	24	B	34	A	44	B
5	B	15	B	*25	A	35	C	45	B
6	B	16	E	26	D	36	D	46	B
7	B	*17	E	*27	D	37	E	47	A
8	B	18	A	28	C	38	E	48	C
9	D	19	B	*29	C	39	A	49	E
10	C	20	C	*30	C	40	C	50	B

Please note that relatively harder questions are indicated by a star (*) preceding the question number.

ANSWERS TO MODEL TESTS

Model Test 9 Answers

#		#		#		#		#	
1	E	11	C	21	E	31	C	41	D
2	D	12	D	22	E	32	D	42	C
3	C	13	D	23	D	33	E	43	A
4	B	14	E	24	E	34	B	*44	D
5	C	15	B	25	B	35	B	45	A
6	E	16	D	26	D	*36	D	46	D
7	B	17	E	27	E	37	C	47	D
8	D	18	A	28	A	38	B	48	B
*9	D	19	B	29	B	*39	A	49	D
10	D	20	D	30	E	*40	E	50	C

Model Test 10 Answers

#		#		#		#		#	
1	B	11	D	21	C	*31	E	41	C
2	D	*12	B	*22	C	*32	B	42	C
3	E	*13	C	23	E	33	C	43	D
4	A	14	D	24	A	34	C	44	C
5	E	15	D	25	A	35	A	45	C
6	A	16	A	26	A	36	E	*46	B
7	D	17	B	27	D	*37	A	47	C
8	C	18	E	28	C	*38	B	48	D
9	E	19	E	29	D	39	D	49	C
*10	B	20	D	*30	D	*40	B	*50	D

Model Test 11 Answers

#		#		#		#		#	
1	D	11	A	21	C	*31	E	41	D
2	C	12	D	22	A	32	A	42	D
3	E	13	C	23	B	33	C	43	D
4	B	14	B	24	C	*34	C	44	B
*5	D	15	A	25	B	35	C	45	B
*6	C	*16	D	*26	B	36	E	*46	E
7	A	17	E	27	E	37	B	47	A
*8	E	*18	C	28	C	38	C	*48	E
9	D	19	D	29	A	39	D	49	D
10	E	20	D	30	B	40	C	*50	C

Model Test 12 Answers

#		#		#		#		#	
1	B	11	B	21	B	31	A	41	D
2	C	12	B	22	C	32	C	42	A
3	E	13	C	23	E	33	B	43	C
4	B	14	D	24	E	34	A	*44	C
5	D	15	E	25	B	35	E	*45	D
6	E	16	D	26	C	36	C	46	A
7	C	*17	D	27	C	37	E	47	C
8	A	18	E	28	A	38	E	48	C
9	C	19	A	29	B	*39	D	*49	E
10	A	20	D	30	D	40	B	50	B

Model Test 13 Answers

#		#		#		#		#	
1	E	11	C	*21	C	31	A	41	D
2	A	*12	C	*22	E	32	C	42	C
3	D	13	A	23	A	33	C	*43	C
4	C	14	B	24	D	34	B	*44	B
5	E	15	D	*25	C	35	A	*45	E
6	A	16	C	26	B	36	D	46	D
7	A	*17	E	*27	E	37	E	47	E
8	E	18	B	28	B	38	A	*48	D
9	C	19	A	29	A	39	C	49	A
*10	D	20	C	30	A	40	C	50	D

Model Test 14 Answers

#		#		#		#		#	
1	D	11	C	*21	D	31	A	41	B
2	E	12	E	22	B	*32	D	42	E
3	E	13	C	23	A	33	D	43	D
4	E	14	E	24	B	34	A	44	B
5	A	15	C	25	B	35	C	45	A
6	E	16	E	26	E	*36	C	46	B
7	D	17	D	27	E	37	D	47	B
8	E	18	A	28	B	38	D	48	B
9	B	19	D	29	A	39	E	49	C
10	D	20	C	30	A	40	C	50	E

Model Test 15 Answers

#		#		#		#		#	
1	E	11	E	21	C	*31	E	41	A
2	C	12	C	22	D	32	C	42	B
3	B	13	A	23	D	33	E	*43	D
4	B	14	D	24	B	34	C	44	E
*5	B	15	D	25	B	35	B	45	B
6	A	16	B	26	C	36	C	46	C
7	B	17	D	27	C	37	A	47	D
8	D	18	B	*28	C	38	A	48	C
9	B	19	A	29	D	39	D	*49	C
10	C	20	B	30	A	40	D	50	E

Please note that relatively harder questions are indicated by a star (*) preceding the question number.

Scaled Score Conversion Table
Mathematics Level 2 Subject Test

Raw Score	Scaled Score	Raw Score	Scaled Score	Raw Score	Scaled Score
50	800	28	650	6	480
49	800	27	640	5	470
48	800	26	630	4	460
47	800	25	630	3	450
46	800	24	620	2	440
45	800	23	610	1	430
44	800	22	600	0	410
43	800	21	590	-1	390
42	790	20	580	-2	370
41	780	19	570	-3	360
40	770	18	560	-4	340
39	760	17	560	-5	340
38	750	16	550	-6	330
37	740	15	540	-7	320
36	730	14	530	-8	320
35	720	13	530	-9	320
34	710	12	520	-10	320
33	700	11	510	-11	310
32	690	10	500	-12	310
31	680	9	500		
30	670	8	490		
29	660	7	480		

Blank Page

Model Test 1 – Solutions

1. (B)

$\log_5 1234 = \dfrac{\log 1234}{\log 5} \cong 4{,}42$

$4 < \log_5 1234 < 5$

2. (A)

{3, 5, 7} is the only prime triple when n is 0 but it is given to be a positive number so there are no prime triples of the given form.

3. (A)

$\left\{\underbrace{2,3,5,7,}_{4}11,13,17,19,23\right\}$; $\dfrac{4}{9}$ of the numbers in this set are less than 8.

4. (E)

The inverse of a constant function is not defined.

5. (D)

$(x-2)(x^2-4)=0$

The zeros are $x_1 = x_2 = 2$ and $x_3 = -2$ and their sum is $2 + 2 - 2 = 2$.

6. (B)

$C = \dfrac{3}{5} + \dfrac{2 \cdot 12}{7} - \dfrac{5}{12} \cdot \left(\dfrac{12}{8}\right)^2 \cong 3.091$ for every square meter. Therefore for 20,000 m^2 it will be $20{,}000 \times 3.091 = 61{,}820$ that is approximately 62,000.

7. (B)

$(\pi, f(\pi)) = (\pi, 2\pi^2 - 2)$

$(g(e), g(0)) = (e^3 - 3e^2 + 3, 3)$

$\text{slope} = \dfrac{2\pi^2 - 2 - 3}{\pi - e^3 + 3e^2 - 3} = 6.63$

8. (C)

$f(2) = \dfrac{3}{2-1} = \dfrac{3}{1} = 3$

$f(1) = 1^2 - 2 \cdot 1 - 3 = 1 - 2 - 3 = -4$

$3 - 4 = -1$

9. (D)

$\dfrac{5 \cdot (x^2 - 1)}{x+1} = \dfrac{5 \cdot (x-1)(x+1)}{(x+1)} = 5x - 5$

and $x \neq -1$ therefore range does not include $5(-1) - 5 = -10$.

10. (D)

	participated the talent show	did not participate the talent show	totals
girls	22	33	55
boys	18	47	65
totals	40	80	120

The probability that a randomly selected student has attended the talent show given that the student is a girl is $\dfrac{22}{55} = \dfrac{2}{5}$.

11. (E)

All integers less than -4 and greater than -2 satisfy the inequality given; therefore there are infinitely many solutions.

12. (D)

By the cosine rule

$81^2 = 45^2 + 72^2 - 2 \cdot 45 \cdot 72 \cdot \text{Cos}\theta$

$-648 = -6480 \text{Cos}\theta \Rightarrow \text{Cos}\theta = \dfrac{1}{10}$

$\theta = \text{Cos}^{-1}\left(\dfrac{1}{10}\right) \cong 84.3°$

13. (C)

The inverse of $f(x) = \sqrt{4x}$ is given by $\dfrac{x^2}{4}$ when x is nonnegative only.

14. (B)

$f(x,y) = 1$ shifted 2 units up and 1 unit toward left to give $f(x+1, y-2)=1$.

15. (E)

$-\dfrac{b}{a}$ and $\dfrac{c}{a}$ are respectively the sum and product of the roots. The sum of the roots is an integer, however the product of the roots is not; therefore $\dfrac{b}{a}$ and $\dfrac{c}{a}$ are both rational but not integral.

16. (B)

Discriminant must be positive therefore $k^2 - 16 > 0$ which implies that $x < -4$ or $k > 4$.

17. (B)

$\dfrac{4}{3}\pi r^3 = \pi r^2 \cdot h \quad \dfrac{4r}{3} = h \quad \dfrac{r}{h} = \dfrac{3}{4}$

18. (D)

Requested volume inside the cylinder and outside the cone given.

$$49\pi \cdot 3 - \frac{49\pi \cdot 3}{3} = 2 \cdot 49\pi = 307.88$$

19. (D)

Sketch the graph of $h(t) = 100 + 20t - 5t^2$ and find the y coordinate of the vertex which is 120.

20. (A)

$$\frac{m+n+p}{2} = p$$

$$m + n + p = 2p$$

$$m + n = p$$

21. (C)

If one root is $4 + \sqrt{2}$ then the other one is $4 - \sqrt{2}$. P the sum of these roots therefore it equals 8.

22. (D)

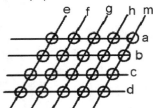

There are 20 intersection points as shown in the figure above.

23. (B)

$$(\sin x + \cos x)^2 = 1.4^2$$

$$\sin^2 x + 2\sin x \cos x + \cos^2 x = 1.96$$

$$1 + 2\sin x \cos x = 1.96$$

$$\sin x \cos x = 0.48$$

24. (B)

$$\tan^{-1}\left(\frac{5.7}{100}\right) = 3.26°$$

25. (E)

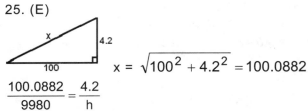

$$x = \sqrt{100^2 + 4.2^2} = 100.0882$$

$$\frac{100.0882}{9980} = \frac{4.2}{h}$$

Therefore road climbs up to a maximum height of h = 418.79 miles that is approximately 419 miles.

26. (A)

The expression is undefined when its denominator is zero.

$\tan^3 x = 0 \Rightarrow \tan x = 0 \Rightarrow x = k\pi$ where k is an arbitrary integer.

27. (E)

The expression 5^{-x} is positive for all values of x.

28. (E)

The expression in E simplifies to

$(x - 3)^2 + (y - 7)^2 = 16$ where (3, 7) is the center and 4 is the radius. Therefore E is the best choice.

29. (C)

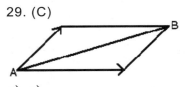

$\vec{u} + \vec{v}$ is the length of the diagonal AB of the parallelogram given above and it can be determined by the cosine rule.

$$x^2 = 49 + 121 - 2 \cdot 7 \cdot 11 \cdot \text{Cos}130°$$

$$x^2 = 268.99$$

$$x = 16.4$$

30. (B)

$$y = x + 7$$

$$x^2 + (x + 7)^2 = 169$$

$$x^2 + x^2 + 14x + 49 = 169$$

$$2x^2 + 14x - 120 = 0$$

$$x^2 + 7x - 60 = 0$$

$$(x - 5)(x + 12) = 0$$

Therefore x is 5 and y is 12 or x is -12 and y is -5.

31. (A)

$$x^9 - 64x^3 = 0$$

$$x^3(x^6 - 64) = 0$$

$$x = 0$$

or

$$x = \pm 2$$

32. (A)

As x becomes very large 5/(7x) becomes zero and f(x) approaches $\frac{3x}{7}$.

33. (D)

$x^{\frac{3}{4}} + 2x - 20 = 0$. Graph the function

$y = x^{\frac{3}{4}} + 2x - 20$ and determine its zeros; one of the zeros is at x = 7.7.

34. (C)

The function is zero at x = 4 + 7k where k is an arbitrary integer. Therefore it has another zero at x = 18.

35. (B)

If the frequency is $\frac{1}{\pi}$ then the period is π. The function given by B has a period that is equal to π.

36. (D)

(a, b) ◊ (c, d) = (a + bc, bd)
(a, b) ◊ (0, 1) = (a, b)
and (0, 1) ◊ (c, d) = (c, d) therefore (0, 1) is the identity element for this operation.

37. (B)

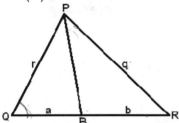

By the sine rule $\dfrac{PB}{\sin Q} = \dfrac{a}{\sin(P/2)}$ and

$\dfrac{PB}{\sin R} = \dfrac{b}{\sin(P/2)}$. Therefore a·sinQ = b·sinR and

$\dfrac{a}{b} = \dfrac{\sin R}{\sin Q}$.

38. (C)

$f(g(x)) = a \cdot g(x) + b = c; g(x) = \dfrac{c-b}{a}$ Therefore g(x) is

a constant function.

39. (E)

As x becomes very large $3e^{-5x}$ becomes zero and f(x) equals 7. Therefore y = 7 is the horizontal asymptote.

40. (E)

The inequality given in E is false for some angle values in the interval
0° < x < 45°.

41. (D)

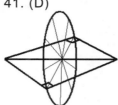

In the three dimensional figure above all such points P constitute a circle whose axis of symmetry coincides with the segment AB.

42. (B)

$g(3) = \dfrac{-10}{3} = -3{,}33$

$g(-6) = \dfrac{-10}{-6} = \dfrac{10}{6} = \dfrac{5}{3} = 1{,}667$

$f(-3{,}33) = -3$

$f(1{,}667) = 2$

$-3 + 2 = -1$

43. (D)

The axis of symmetry of h(x) has to be the line given by x = 2. Moreover sum of the zeros is 2 · 2.

44. (E)

All quantities can be determined.

45. (B)

$\log \dfrac{1}{2} + \log \dfrac{2}{3} + \log \dfrac{3}{4} + \dots + \log \dfrac{99}{100}$

$= \log \dfrac{1}{2} \cdot \dfrac{2}{3} \cdot \dfrac{3}{4} \dots \dfrac{99}{100} = \log \dfrac{1}{100}$

$= \log 10^{-2} = -2$

46. (C)

The given point is in the second quadrant; its magnitude is r and the angle it makes with the positive x axis is θ that can be determined in the following way:

$\theta = \pi - \dfrac{\pi}{4} = \dfrac{3\pi}{4}$

$r^2 = 2^2 + 2^2 \Rightarrow r^2 = 8 \Rightarrow r = 2\sqrt{2}$

47. (C)

In the worst case Defne picks 1 of each color first. The next one that she picks has to match the color of one of the already picked socks.

48. (E)

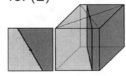

There are infinitely many planes that satisfy all three conditions. The figure to the left above is the upper view of the cube given to the right.

49. (D)

$f(3) = f(2) \cdot f(1) = 1 \cdot 2 = 2$

$f(4) = f(3) \cdot f(2) = 2 \cdot 2 = 4 = 2^2$

$f(5) = f(4) \cdot f(3) = 2 \cdot 2 \cdot 2 = 8 = 2^3$

$f(6) = f(5) \cdot f(4) = 2^3 \cdot 2^2 = 25$

$f(7) = f(6) \cdot f(5) = 2^5 \cdot 2^3 = 2^8$

$f(8) = f(7) \cdot f(6) = 2^8 \cdot 2^5 = 2^{13}$

$f(9) = f(8) \cdot f(7) = 2^{13} \cdot 2^8 = 2^{21}$

50. (D)

$\begin{vmatrix} 2 & 4 & -1 \\ 1 & -2 & 5 \\ 1 & 1 & 0 \end{vmatrix} = 7$. The calculation can be carried out via any scientific or graphing calculator such as TI 83, 84, or 89.

Model Test 2 – Solutions

1. (A)

$y^2 = 16 \Rightarrow y = \mp 4$

$x = \sqrt{9} = 3$

So, $x + y = 3 \pm 4$; 1 or -7.

2. (C)

The following equality is valid when n is very large: $e = \left(1 + \dfrac{1}{n}\right)^n$

$\left(1 + \dfrac{1}{100000}\right)^{10000} = \left[\left(1 + \dfrac{1}{100000}\right)^{100000}\right]^{0.1}$

$= e^{0.1}$

3. (C)

$4 \cdot (w - 3) = (w + 1) \cdot (w - 5)$

$\Rightarrow 4w - 12 = w^2 - 4w - 5$

$\Rightarrow w^2 - 8w + 7 = 0$

$\Rightarrow (w - 7)(w - 1) = 0$

Therefore w=1 or w=7

4. (D)

$8 = 4 \cdot \ln(t)$

$\Rightarrow \ln(t) = 2$

$\Rightarrow t = e^2$

$b = 3 \cdot \sin(e^2) \cong 2.681$

5. (A)

p is a multiple of 10 and only I must be correct.

6. (C)

$f^{-1}(f(x)) = x$ as long as $f^{-1}(x)$ exists thus $f^{-1}(f(3)) = 3$.

7. (A)

If n is a divisor of 60, so is −n; therefore all divisors of 60 add up to 0. So the sum of the elements in set A is 0.

8. (C)

Let ab be the two digit positive integer; then ab + 45 = ba

$\Rightarrow 10a+b+45=10b+a \Rightarrow 9(b-a)=45$

$\Rightarrow b-a=5 \Rightarrow b=a+5$

All possibilities are given in the following table.

Digit a	Digit b	Two digit number ab
1	6	16
2	7	27
3	8	38
4	9	49

9. (A)

For m = 0, y = 1 and the graph of y = 1 is a line parallel to the x-axis. And it has no points in the third and forth quadrants.

10. (E)

Plot the graph of $y = 2^{2x} + 2^{-2x} - 5$ and find its zeros that are ± 0.88.

11. (C)

Let x_1 and x_2 be the roots of the given quadratic equation. Then,

$x_1 = 3 + \sqrt{2}$

$x_2 = 3 - \sqrt{2}$

$x_1 \cdot x_2 = Q = (3 + \sqrt{2}) \cdot (3 - \sqrt{2}) = 9 - 2 = 7$

12. (B)

If p<q and $\dfrac{1}{p} < \dfrac{1}{q}$ then p<0<q; therefore pq<0.

13. (E)

Let (x, y) be any point on the coordinate plane. (x, -y) is the point symmetric to (x,y) with respect to the x-axis.

14. (D)

For any quadratic equation with real coefficients, if a + bi is one root, the other one is a − bi or vice versa.

15. (A)

Plot the curve $y = \ln x - e^{-2x}$ with the graphing calculator and find its zero (x intercept) and the answer is 1.1138.

16. (A)
$$\cos^2(2x) + \sin^2(2x) = 1$$
$$\Rightarrow \cos^2(2x) - 1 = -\sin^2(2x)$$

17. (B)
The points that satisfy the given inequality are those on the lines y = ± x and in between.

18. (C)
It is required that n be an integer greater than 7 and the statement given in C is false because 7 is a positive multiple of 7 and it does not meet the requirement we have.

19. (C)
The function $y = \sqrt{4x-3} + \sqrt{8x+1} - 8$ intersects the x axis at a single point.

20. (E)
$$-1 \le \cos\theta \le 1 \Rightarrow -4 \le 4 \cdot \cos\theta \le 4$$

21. (B)
AB = OC'=y and the angle that OC' makes with the vertical is θ. Using right angle trigonometry correct answer is given in B.

22. (D)
$$\left[2\left(x - \frac{3}{2}\right)\right]^2 - \left[5\left(y - \frac{4}{5}\right)\right]^2 = 6$$
$$4\left(x - \frac{3}{2}\right)^2 - 25\left(y - \frac{4}{5}\right)^2 = 6$$
$$\Rightarrow \frac{\left(x - \frac{3}{2}\right)^2}{\frac{3}{2}} - \frac{\left(y - \frac{4}{5}\right)^2}{\frac{6}{25}} = 1$$

It's a hyperbola centered at $\left(\frac{3}{2}, \frac{4}{5}\right)$.

23. (B)
$$f \circ g \circ g^{-1} = f$$
$$g^{-1}(x) = \frac{x+1}{2}$$
$$f(x) = 3\left(\frac{x+1}{2}\right) + 4 = \frac{3x+11}{2}$$
Therefore $f^{-1}(x) = \frac{2x-11}{3}$

24. (C)
This is a tricky question. Suppose f(x) and g(x) are given as follows.

$$f(x) = \frac{1}{x-1}$$
$$g(x) = \frac{1}{x+1}$$
$$f(g(x)) = \frac{1}{\frac{1}{x+1} - 1} = \frac{x+1}{-x}$$

f(x) is defined for all real numbers other than 1 and g(x) is defined for all real numbers other than -1. However f(g(x)) is not defined for both 0 and -1.

25. (C)
The composition of two identical functions gives a quadratic function therefore each identical function has to be quadratic.

26. (A)
Let x be one side of the cube.
So, $x^3 = 216 \Rightarrow x = 6$
The new object is a regular octahedron that is two square pyramids whose bases coincide and the volume of this solid is: $2 \cdot \left(3\sqrt{2}\right)^2 \cdot 3 \cdot \frac{1}{3} = 36$.

27. (E)
Range of f(x) is y > 0 or y ≤ -1 and not all real numbers.

28. (D)
$$c = 2R\cos x$$
$$a = 2R\sin\left(\frac{\alpha}{2}\right)$$
$$b = 2R\cos\left(\frac{\alpha}{2} - x\right)$$
Perimeter = a + b + c =
$$2R\left(\sin\frac{\alpha}{2} + \cos x + \cos\left(\frac{\alpha}{2} - x\right)\right)$$

29. (B)
The point A and B are the foci of the ellipse.
Since $|AP_1| + |P_1B| = |AP_2| + |P_2B|$, the shape will be an ellipse.

30. (C)
$$\frac{f(a+4)}{f(a+3)} = \frac{(a+4) \cdot 2^{a+4}}{(a+3) \cdot 2^{a+3}} = \frac{2(a+4)}{a+3} = 6$$
$$\Rightarrow 3(a+3) = a+4 \Rightarrow 3a+9 = a+4$$
$$\Rightarrow 2a = -5 \Rightarrow a = -\frac{5}{2}$$

31. (B)
The only solution requires that y be 0 and cos(2x) be 1. Therefore x can be 0 or π only.

32. (E)

$\sin x \cdot \dfrac{1}{\sin x} = 1$ when $\sin x$ is nonzero. Therefore x can be any real number other than the integer multiples of π

33. (E)

The graph of $y = \cos(3x) + x^2 - \dfrac{1}{2}$ intersects the x axis 4 times; make sure that you plot this function in the radian mode.

34. (B)

In order to determine the equation of a parabola of the form $y = ax^2 + bx + c$ you need either three points none of which is the vertex or two points one of which is the vertex.

35. (E)

The graph of $g(x) = f(x+1) - 1$ looks like:

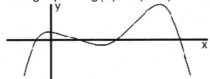

And it intersects the x axis 4 times.

36. (B)

The domain is those (x, y) points for which $x^2 + y^2 - 1$ is greater than or equal to zero;

$\Rightarrow x^2 + y^2 - 1 \geq 0 \Rightarrow x^2 + y^2 \geq 1$ that are all points on and outside the unit circle.

37. (A)

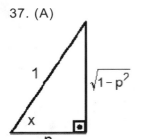

The cosine and the tangent functions both have the same sign in the given interval. So,

$\tan x = \dfrac{\sqrt{1-p^2}}{p}$

38. (D)

Let x be the angle that OB makes with the positive x axis; then

$\sin x = 0.6 \Rightarrow \alpha \cong 37°$. The measure of angle AOB is $2 \cdot \left(90° - \alpha\right) \cong 106°$.

39. (C)

It takes Nuran 4 hours to arrive in Bodrum and it takes Nazan 4.2 hours to arrive in Bodrum. 4.2 hours means 4 hours and 12 minutes. Therefore Nazan will arrive in Bodrum at 9:12 AM.

40. (C)

The set is $\{1,10,19,28,37,46,\ldots,199\}$ and it has 23 elements.

41. (C)

	X	X	X	
X				X
X				X
X				X
X				X
X				X
	X	X	X	

There are $5 \cdot 7 = 35$ squares. There are 16 squares each of which having 5 neighbors (denoted by x). So, the probability to select one of them is 16/35.

42. (B)

$80 - 5t^2 = 0 \Rightarrow t^2 = 16 \Rightarrow t = 4$

for $t = 4$

$x = 30 \cdot t = 120$

43. (B)

$\dfrac{1}{f(x)}$ is zero when $f(x)$ is negative or positive infinity; that is around the vertical asymptote.

Therefore $\dfrac{1}{f(x)}$ has a zero at x = 2.

44. (A)

$8 \Delta 9 = 2^3 \Delta 3^2 = 2 \cdot 3 + 3 \cdot 2 + 4 = 16$

45. (E)

Domain has two elements only therefore the range cannot have more than two elements.

46. (D)

In order to have two complex conjugate zeros, the quadratic function must have coefficients that are all real or all complex therefore D is false.

47. (D)

$|x| \cdot \left(|x|^2 + 2 \cdot |x| - 3\right) = |x| \cdot \left(|x| + 3\right) \cdot \left(|x| - 1\right) = 0$

Therefore x is 0 or ±1.

48. (C)

Area of the shaded region is $7 \cdot 4 \cdot \sin(20°)/2 = 4.79$ square inches.

49. (B)

1.5 miles is 2640 yards. So the height of the building above the top of the tree is $2640 \cdot \tan(0.03) = 79.22$ yards. 15 feet equals 5 yards so the total height is $79.22 + 5 \cong 84.22$ yards.

50. (E)

Plot the graph of $y = |\tan(3x)|$ in the radian mode to see that its period is $\pi/3$.

Model Test 3 – Solutions

1. (B)

$$34^x = 14 \cdot \frac{134}{100} = 18.76$$

$$\log 34^x = \log 18.76$$

$$x \cdot \log 34 = \log 18 \cdot 76$$

$$x = \frac{\log 18.76}{\log 34}$$

$$x = 0.831$$

2. (E)

$$\sqrt[3]{x-4} = 5$$

$$x - 4 = 5^3 = 125$$

$$x = 129$$

3. (A)

f(x) is the inverse of function g(x)

g(x)=2x+1

y=2x+1

x=2y+1

$$y = \frac{x-1}{2} \Rightarrow f(x) = \frac{x-1}{2}$$

4. (C)

$$a = 0.2 \Rightarrow b = 37 \cdot 1.37^{0.2} = 39.40$$

$$c = (3\pi)^{0.25} \cdot \left(\frac{39.40}{2}\right)^3 = 13395.73$$

$$d = (2 \cdot 13395.73 - 39.40)^{0.5} = 163.56 \approx 164$$

5. (C)

f(x) = mx + b

$$f(2) = 5 \Rightarrow 5 = 2m + b$$

$$f(-1) = 2 \Rightarrow 2 = -m + b$$

m = 1

b = 3

$$\Rightarrow f(x) = 1x + 3$$

$$f(x) = x + 3$$

$$f(-3) - 3 + 3 = 0$$

6. (D)

$$\frac{10.80 - 40.5}{5} = \frac{800 - 200}{5} = \frac{600}{5} = 120$$

7. (D)

f(a)=9 therefore a²-5=9, a²=14 and

$$a = \pm\sqrt{14} = \pm 3.74$$

8. (A)

$$d = \sqrt{(4-1)^2 + (5--2)^2 + (-2--6)^2}$$

$$= \sqrt{9 + 49 + 16} = \sqrt{74}$$

9. (D)

$$\text{new cost per person} = \frac{\text{total cost}}{\text{new number of people}}$$

$$= \frac{12.50 \cdot 20}{20 + n}$$

10. (A)

x⁵⁰(x+3)=k²

x+3 must be a perfect square

⇒x can be 22.

11. (B)

a₁=5=a

a₂=7=a+d

⇒d=2

aₙ=a+(n-1).d

5+(n-1).2>2000

2n-2+5>2000

2n>1997

$$n > \frac{1997}{2} = 998.5$$

$$\Rightarrow \min(n) = 999$$

12. (D)

$$\csc(x) = \frac{1}{\sin(x)} \Rightarrow b = \frac{1}{a}$$

13. (D)

cosx decreases throughout in the first quadrant

and $\left(\frac{\pi}{3}, \frac{\pi}{2}\right)$ is in the first quadrant.

14. (E)

$$\sin x = 35° = \frac{35}{180} \cdot \pi = 0.610$$

$$\Rightarrow x = \sin^{-1} 0.610 = 0.657$$

$$x = 0.657 + k \cdot 2\pi$$

or

$$x = \pi - 0.657 + k \cdot 2\pi$$

where k is an integer. Therefore x cannot be 0.574.

15. (E)

2c-1 is odd there fore ab is odd

⇒a and b most be both odd.

16. (B)

f(0)= -5 and f(5)=0.

The function in II satisfies both.

17. (E)

$$\text{mean} = \frac{\text{sum}}{\text{number of terms}}$$

$$\Rightarrow \text{sum} = \text{mean} \cdot \text{number of terms}$$

$$\Rightarrow \text{mean} = \text{median}.$$

18. (A)
Take x=1 and y=-1 for instance.
x-y=1- -1=2>0

19. (D)
$x^2-9>0$

$x^2>9$ therefore $\sqrt{x^2} > \sqrt{9}$ and

$|x| > 3 \Rightarrow x > 3\, or\, x < -3$

20. (B)
$x \le 0 \Rightarrow$ B,C,D,E
$x \le y \Rightarrow$ A,B
$y \le -x \Rightarrow$ A,B,C,D $\}$ B satisfies all

21. (D)

$$\frac{3m-1+a}{5m+b} = \frac{3\left(m+\frac{a-1}{6}\right)}{5\left(m+\frac{6}{5}\right)} \Rightarrow \frac{a-1}{3} = \frac{6}{5}$$

$$\Rightarrow 18 = 5a-5 \Rightarrow a = \frac{23}{5}$$

22. (D)

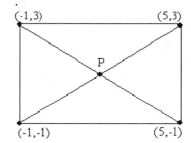

Line m must pass through P.

$$P = \left(\frac{5-1}{2}, \frac{3-1}{2}\right) = (2,1)$$

$$1 = 2+c \Rightarrow c = -1$$

23. (C)
Amplitude $= \dfrac{y\max-y\min}{2}$

If f(x) is plotted then it will be seen that y_{max}=2 and y_{min}= -2.

$\Rightarrow$ Amplitude $= \dfrac{2--2}{2} = \dfrac{4}{2} = 2$

24. (E)
$a_1 = 5 = a$
$a_2 = 7 = a.r$

$\Rightarrow r = \dfrac{7}{5} = 14$

$$5_{30} = \frac{a\left(1-r^{30}\right)}{1-r} = \frac{5.\left(1-1.4^{30}\right)}{1-1.4} = 302,505$$

25. (E)
$a \ne 0$ otherwise f(x) is not quadratic and can never have two distinct zeros.

26. (E)
Two different colors:
BR,RB,BG,GB,RG,GR

$\Rightarrow$ Probability $= \dfrac{7}{24} \cdot \dfrac{8}{23} \cdot 2 + \dfrac{7}{24} \cdot \dfrac{9}{23} \cdot 2 + \dfrac{8}{24} \cdot \dfrac{9}{23} \cdot 2$

without replacement $\leftarrow$

different orderings $\leftarrow$

27. (C)
$5 = \log_a(2+b) \Rightarrow 2+b = a^5$
$2 = \log_a(-1+b) \Rightarrow b-1 = a^2$

$a^5 - 2 = a^2 + 1 \Rightarrow a^5 - a^2 - 3 = 0$
Plot the graph of $y=x^5-x^2-3$ and find its x intercept.
$\Rightarrow x = 1.3734 = a$
$\Rightarrow b = a^2 +1 = 2.886 \Rightarrow b = 2.89$

28. (E)
f(x+1)+1

$\rightarrow$ shift f(x) for 1 unit up.
$\rightarrow$ shift f(x)for 1 unit left

29. (E)
$\log\left(\dfrac{a}{b}\right) = \log(a) - \log(b)$

$\Rightarrow f(x) = \log x$

30. (B)
$x = \sqrt{4^2 + 6^2} = 7.21$
(Pythagoras' Theorem)

$\tan 56° = \dfrac{x}{h} \Rightarrow h = \dfrac{x}{\tan 56°} = \dfrac{7.21}{\tan 56°} = 4.864$

volume $= 6 \cdot 4 \cdot 4.864 = 116.73 \approx 117$

31. (B)
$$\begin{bmatrix} 1 & 2 & -3 \\ 2 & 1 & 0 \\ -1 & 1 & 1 \end{bmatrix}\begin{bmatrix} x \\ y \\ z \end{bmatrix} = \begin{bmatrix} 3 \\ 4 \\ -3 \end{bmatrix} \Rightarrow \begin{bmatrix} x+2y-3z \\ 2x+y \\ -x+t+z \end{bmatrix} = \begin{bmatrix} 3 \\ 4 \\ 3 \end{bmatrix}$$
$x + 2y + 3z = 3$
$2x + y = 4$
$-x + y + z = -3$

32. (D)

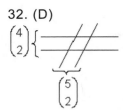

When two pairs of parallel lines intersect, one parallelogram forms ⇒ number of

parallelograms is $\binom{4}{2}\cdot\binom{5}{2} = 6\cdot 10 = 60$.

33. (D)

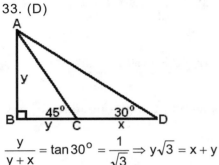

$$\frac{y}{y+x} = \tan 30° = \frac{1}{\sqrt{3}} \Rightarrow y\sqrt{3} = x+y$$

$$y = \frac{x}{\sqrt{3}-1}$$

34. (A)

$4^x - x > 3$; so $4^x - x - 3 > 0$. Plot the graph of $y = 4^x - x - 3$

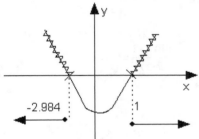

35. (B)

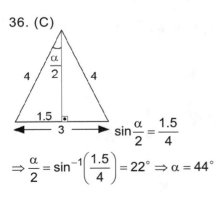

$\underline{9}\ \underline{9}\ \underline{9}\ \underline{9}\ \underline{9}$

cannot be the same as the immediate left hand neighbor.

0 cannot be the leftmost digit.

36. (C)

$\sin\frac{\alpha}{2} = \frac{1.5}{4}$

$\Rightarrow \frac{\alpha}{2} = \sin^{-1}\left(\frac{1.5}{4}\right) = 22° \Rightarrow \alpha = 44°$

37. (D)

$$\frac{V\,\text{upper cone}}{V\,\text{bigger cone}} = \frac{1}{2} = k^3 \Rightarrow \text{side ratio} = \frac{1}{\sqrt[3]{2}} = k$$

$$\Rightarrow \text{area ratio} = k^2 = \frac{1}{\sqrt[3]{4}}$$

38. (E)

$$\sin\theta = \frac{x}{R} \Rightarrow R = \frac{x}{\sin\theta}$$

$$\tan\theta = \frac{PQ}{2R}$$

$$PQ = 2R\tan\theta$$

$$= 2\cdot\frac{x}{\sin\theta}\cdot\frac{\sin\theta}{\cos\theta}$$

$$= \frac{2x}{\cos\theta}$$

39. (C)

period of $f(x) = \frac{2\pi}{\pi} = 2$

frequency of $f(x) = \frac{1}{\text{period}} = \frac{1}{2}$

40. (C)

statement: $p \Rightarrow q$

inverse: not $p \Rightarrow$ not q

41. (E)

$$\left(x - \frac{1}{x}\right)^2 = 4^2 \Rightarrow x^2 - 2 + \frac{1}{x^2} = 16 \Rightarrow x^2 + \frac{1}{x^2} = 18$$

42. (C)

Inequality in C does not represent the shaded region.

43. (C)

y must be 0 therefore $\sin(2x) = 1$

$$2x = \frac{\pi}{2} \Rightarrow x = \frac{\pi}{4}$$

44. (C)

$|x| = -x \Rightarrow x$ is negative or zero

45.

$$\alpha = \tan^{-1}\left(\frac{15}{7}\right) = 64.98°$$

$$\beta = \tan^{-1}\left(\frac{3}{7}\right) = 23.20°$$

$$\beta + x = \alpha \Rightarrow x = \alpha - \beta = 41.78° \approx 42°$$

46. (C)

$x = 1 + 2t$

$y = 3 + 4t$

Eliminate t and you get

$4t = y-3 = 2(x-1)$

$y-3 = 2x-2$

$y = 2x+1 \Rightarrow$ slope = 2

47. (C)

$4x^2 - 2kx + 1 = k - 2$

$4x^2 - 2kx + 3 - k = 0$

$\Delta > 0 \Rightarrow (2k)^2 - 4.4(3-k) > 0$

$4k^2 - 48 + 16k > 0$

$k^2 - 12 + 4k > 0$

$k^2 + 4k - 12 > 0$

$\Rightarrow k < -6 \text{ or } k > 2$

48. (A)

Side of the square base = x

$x = R\sqrt{2} = 6\sqrt{2}$

Volume $= x^2 \cdot \dfrac{h}{3} = \dfrac{\left(6\sqrt{2}\right)^2 \cdot 6}{3} = 144$

49. (A)

$\lim\limits_{x \to -1} \dfrac{x^2 - 1}{x+1} = \lim\limits_{x \to -1} \dfrac{(x-1)(x+1)}{x+1} = -1-1 = -2$

50. (D)

I. $g(-x) = f^2(-x)$ need not equal $f^2(x)$

II. $g(-x) = f(-(-x)) + f(-x) = f(x) + f(-x) = g(x)$

III. $g(-x) = f(-|-x|) = f(-|x|) = g(x)$

$\Rightarrow$ Answer is II and III only

Model Test 4 – Solutions

1. (A)

$x = \sqrt{e^{\ln 3} - \sqrt{3}} = 1.126$

2. (D)

Add all equations side by side and you get

$2x + 2y + 2z = 13 + 14 + 15$

$2x + 2y + 2z = 42$

$x + y + z = 21$

3. (A)

$\dfrac{2^x}{2^y} = \dfrac{5}{8} \Rightarrow 2^{x-y} = \dfrac{5}{8}$

$x - y = \log_2(5/8) = \dfrac{\log(5/8)}{\log 2} \cong -0.68$

4. (E)

$-9 < 3x - 1 < 9$

$-8 < 3x < 10$

$-2.67 < x < 3.33$

$x \in \{2, -1, 0, 1, 2, 3\}$

5. (D)

$\dfrac{2.5}{2\pi} = \dfrac{D}{360°} \Rightarrow D = \dfrac{360° \cdot 2.5}{2\pi} \cong 143°$

6. (E)

It can be seen that the points A through D are on the line y = 2x + 3 but E is not.

7. (D)

$7 \cdot 14 \cdot 21 \cdot 28 \cdot 35 \cdot 42 \cdot 49 = 4,150,656,720$

8. (C)

The expression given in C is the negative of all others.

9. (B)

$e, \ln 2, \sqrt[3]{5}, \sqrt{3}, \pi$ are irrational numbers.

10. (C)

$C(7,2) = \dfrac{7!}{5! \cdot 2!} = 21$

11. (D)

$\dfrac{\sin x \cdot \left(\sin^2 x + \cos^2 x\right)}{\cos x} = \dfrac{\sin x \cdot 1}{\cos x} \Rightarrow \tan x = 2.5$

$\Rightarrow x = \tan^{-1} 2.5 = 68.2° \cong 68°$

12. (C)

$75 = 100 + 20t - 5t^2 \Rightarrow 5t^2 - 20t - 25 = 0$

$t^2 - 4t - 5 = 0 \Rightarrow (t-5)(t+1) = 0 \Rightarrow t = 5$

13. (D)

Original period is multiplied by the reciprocal of the coefficient of x to get the new period.

14. (B)

PD + QD = length of the major axis = 10

15. (C)

$g(3) = \dfrac{-10}{3} = -3.33 \Rightarrow f(-3.33) = -4$

$g(-6) = \dfrac{10}{6} = 1.67 \Rightarrow f(1.67) = 1$

$-4 + 1 = -3$

16. (A)

$t_1 = 2, t_2 = 5, t_3 = 10, t_4 = 17$

$t_n = n^2 + 1$

17. (D)
f(-1) and f(3) both equal 11.

18. (D)
$(x,y) \xrightarrow{\text{origin}} (-x,y)$
$-y = -x^3 - x^2 \Rightarrow y = x^3 + x^2 \Rightarrow f(x) = x^3 + x^2$

19. (D)
$f(y) = a(y-3)(y-1)$
$f(0) = 1 = a \cdot (-3) \cdot (-1) = 1 \Rightarrow a = \frac{1}{3}$
$f(y) = \frac{1}{3}(y-3)(y-1) = \frac{1}{3}(y^2 - 4y + 3)$

20. (B)
$f(g(x)) = 2 \cdot g(x) - 1 = 3x + 4$
$g(x) = \frac{3x+5}{2} \Rightarrow g^{-1}(x) = \frac{2x-5}{3}$

21. (C)

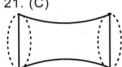

22. (C)
There are three possible cases OOOEE, EOOOE, and EEOOO. Therefore the requested probability equals: $0.6^3 \cdot 0.4^2 \cdot 3$.

23. (D)
$\frac{f-1}{2} = m \Rightarrow 2m = f - 1$

24. (E)
Sketch $y = -3 \cdot \sin x + 4 \cdot \cos x + 1$
and Amplitude $= \frac{y_{max} - y_{min}}{2} = \frac{6 - -4}{2} = 5$

25. (E)
For maximum value x = -1 and y = -3.
$(-1-3)^2 = (-4)^2 = 16$

26. (C)
$2^2 \cdot 3^3 \cdot 5 = 4 \cdot 27 \cdot 5 = 540$
$(1 + 2 + 2^2) \cdot (1 + 3 + 3^2 + 3^3) \cdot (1 + 5)$
The above expression generates the sum of all positive distinct divisors of 540.

27. (C)
$2\cos\left(x + \frac{\pi}{2}\right) = 2 \cdot (-\sin x) = -2\sin x$

28. (E)
$x = a(y-h)^2 + k$ where vertex is given by (k, h).

29. (D)
B and D are not sufficient to determine the equation of the parabola.

30. (C)
3, 4, and 5 are always false.

31. (D)
I, II, IV, V are possible but III is not.

32. (D)
Sketch the graph of
$y = \sin\left(x + \frac{\pi}{2}\right) - \cos(x - \pi)$ in the interval
$0 < x < 2\pi$ and the x values for which y is
positive will be $0 < x < \frac{\pi}{2}$ or $\frac{3\pi}{2} < x < 2\pi$.

33. (D)
Surface Area $= 2(6 \cdot 8 + 6 \cdot k + 8 \cdot k) = 376$
$\Rightarrow 48 + 14k = 188 \Rightarrow k = 10$
Volume of the rectangular box $= 6 \cdot 8 \cdot 10 = 480 = a^3$
$a = 7.82$ feet $\cong 94$ inches

34 (D)
$x^{1.2} = 3; x = 3^{\frac{1}{1.2}} = 2.5$ and $y = 4^{5.6}; y = 2352.5$
$\log_x y = \frac{\log 2352.5}{\log 2.5} = 8.48$

35. (B)
for x<-3 f(x) increases throughout

36. (E)
$\frac{f(x)}{g(x)}$ does not intersect the x axis at all.

37. (C)
The coordinates of point C are given by (2+8/2, 9+4/2) = (6, 11)

38. (E)
When x increases without bound, f(x) increases in a bounded fashion, it always stays less than 0.

39. (E)

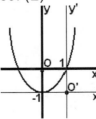

With respect to the x' − y' axes correct answer is given in D.

40. (A)
$\sin(\pi - x) = \sin x \Rightarrow f(x) = \sin x$

41. (E)

P(x) has 4 zeros So $\dfrac{1}{P(x)}$ has 4 vertical asymptotes.

42. (C)

f(x) can be periodic, piecewise and the greatest integer function.

43. (E)

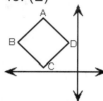

Center of the square is at (-4, 4) and center to vertex distance is 3.

44. (D)

In one year the author sleeps 17 hours per day for at most 5 times and 3 hours per day in the rest of the days.

Therefore he sleeps for at most
$5 \cdot 17 + 360 \cdot 3 = 1165$ hours in a year.

45. (D)

The relative difference between the terms is the greatest in D therefore the standard deviation is the greatest for D.

46. (B)

$x = 1 \Rightarrow y = 3 \cdot \sin\left(1 + \dfrac{\pi}{4}\right) = 2.93$

47. (D)

$\lim\limits_{x \to 1} \dfrac{x^3 - 1}{\ln x}$ can be calculated by replacing x by

0.9999 and 1.0001 in the expression $\dfrac{x^3 - 1}{\ln x}$.

Each calculation gives a value very close to 3. Therefore limit is 3.

48. (E)

The matrix multiplication given by **PS** cannot be carried out since the number of columns in **P** does not equal the number of rows in **S**.

49. (D)

Experimenting with selected x and y values will show that c must satisfy the relation 0 < x < 1.

50. (C)

The number printed is the integer power of $\dfrac{1}{3}$

that is less than 0.000001 for the first time; that is the smallest integer that satisfies the relation

$\left(\dfrac{1}{3}\right)^N < 0.000001$ and it is 13.

Model Test 5 – Solutions

1. (C)
$1.208^7 = 3.754$

2. (C)

$p \cdot \sqrt[3]{p^2} = p^{\frac{1}{2}} \cdot p^{\frac{2}{3}} = p^{\frac{7}{6}} = \sqrt[6]{p^7} = p\sqrt[6]{p}$

3. (C)

Amplitude = $\dfrac{y_{max} - y_{min}}{2}$

If f(x) is graphed, $y_{max} = 4$ and $y_{min} = 1$ will be seen.

Amplitude is $\dfrac{4 - 1}{2} = \dfrac{3}{2} = 1.5$.

4. (E)

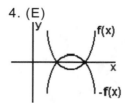

f(x) must intersect the x axis at two distinct points. Therefore discriminant must be positive.

5. (B)
y = sinx and y = cosx can be graphed with the following settings:

$\begin{cases} x_{min} = 0 \\ x_{max} = 2\pi \\ x_{scl} = \pi/2 \end{cases}$ angle mode: radians.

Both graphs decrease in the 2nd quadrant.

6. (B)

$\dfrac{16y^2 - 9x^2}{144} = 1 \Rightarrow \dfrac{y^2}{9} - \dfrac{x^2}{16} = 1$

The graph of the above relation represents a y-hyperbola whose center is at the origin.

7. (B)
If 4 + 2i is a zero then so is 4 – 2i.

$S = Sum = 4 + 2i + 4 - 2i = 8$

$P = Product = (4 + 2i) \cdot (4 - 2i) = 16 + 4 = 20$

$\Rightarrow P(x) = x^2 - Sx + P = x^2 - 8x + 20$

8. (B)
The only possibility is z – 1 = y + 3 = 0 therefore z = 1 and y = – 3 which implies that y + z = – 2.

9. (C)
The figure shows the side view of the picture. The bold lines denote the given planes. The locus of points at a distance of 3 inches from each plane is two parallel planes denoted by the thinner lines. The intersection of two planes gives a line denoted by a point. The result is four parallel planes.

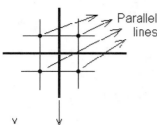

Parallel lines

Perpendicular planes

10. (E)
Least common multiple of 4, 5, 6 is 60. 1020 is the least multiple of 60 greater than 1000. Answer is 1020 + 3 = 1023.

11. (D)

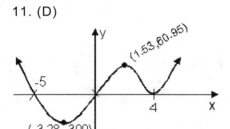

(1.53, 60.95)

-5

4

(-3.28, -300)

For a function to be invertible is must be one to one and onto. For -2 < x < 1, f(x) is invertible.

12. (E)
$\tan^{-1}\dfrac{12}{5} = 67.4°$
$x = 180° + 67.4° = 247.4°$
$\sin x = -0.9231 = -\dfrac{12}{13}$

13. (D)
$\left.\begin{array}{l} 11 \le 2x + 5 \le 11 \\ 2x + 5 < 0 \end{array}\right\}$
$-11 \le 2x + 5 < 0$
$-16 \le 2x < -5$
$-8 \le x < -2.5$

14. (B)
The point (72.5°, 0.3007) is the midpoint of the line segment whose endpoints are (72°, 0.3090) and (73°, 0.2934). Therefore cosx has been assumed to be linear in this interval.

15. (D)
I is equivalent to the statement itself and III is the contrapositive of the statement. Both of I and III represent the same statement.

16. (A)
By the sine rule:

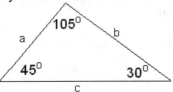

105°

a

b

45°

30°

c

$\dfrac{c}{a} = \dfrac{\sin 105°}{\sin 30°} = 1.93185 = \dfrac{\sqrt{3}+1}{2} \cdot \sqrt{2}$

17. (E)
$r = \dfrac{4}{\cos\theta} \Rightarrow r\cos\theta = 4 \Rightarrow x = 4$

x = 4 is the equation of a line.

18. (C)
Correct ordering is given in C.

19. (E)
$\left.\begin{array}{l} a^{x+3} = a^4 \\ a^{5y} = a^{20} \end{array}\right\}$ If a is not 0, 1 or -1 then x + 3 = 4
and 5y = 20 will be valid. However if a is 0, 1 or -1 then there can be multiple solutions for x and y. Answer is given in E.

20. (C)
$\tan x + \cot x = \dfrac{\sin x}{\cos x} + \dfrac{\cos x}{\sin x} = \dfrac{\sin^2 x + \cos^2 x}{\sin x \cos x}$
$= \dfrac{2 \cdot 1}{2 \cdot \sin x \cos x} = \dfrac{2}{\sin(2x)} = 2\csc(2x)$

21. (A)
$\cot^{-1}\left(\dfrac{-3}{4}\right) = \dfrac{\pi}{2} - \tan^{-1}\left(\dfrac{-3}{4}\right) = 2.2143$
$\sin(2.2143) = 0.8$

22. (B)
$16(x^2-2x+1)+9(y^2+4y+4)=92+16+36=144$
$\dfrac{(x-1)^2}{9} + \dfrac{(y+2)^2}{16} = 1$ is the equation of an ellipse
with semi-minor and semi-major axis lengths of 3 and 4. Therefore area of the ellipse is $\pi \cdot 3 \cdot 4 = 12\pi$.

23. (A)
If E=2 then the third equation is a multiple of the second one and thus the system will have infinitely many solutions.

24. (E)
$\left.\begin{array}{l} e^{-t} = 2 - x \\ e^t = y - 1 \end{array}\right\}$
$(2-x)\cdot(y-1)=1 \Rightarrow y - 1 = \dfrac{1}{2-x}$

25. (D)
$f(x) \to f(2x) \to f(2x+6) \to -f(2x+6) \to 1-f(2x+6)$
$1 \to 5 \to 3 \to 4 \to 2$

26. (D)
$\dfrac{A}{B} = \dfrac{5}{8}$ and $\dfrac{B}{C} = \dfrac{6}{11}$. If B = 48 then A = 30 and C
= 88. $\Rightarrow$ A+B+C = 166 pounds = 2656 ounces

27. (B)

$$f(x) \to f(-x) \to |f(-x)| \to -|f(-x)|$$

reflect the original curve in y; reflect the positive y portion of the resulting curve in x.

28. (C)

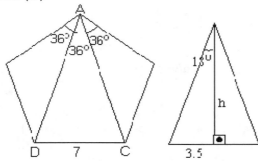

$$\tan 18° = \frac{3.5}{h} \text{ therefore } h = \frac{3.5}{\tan 18°} = 10.77$$

And area of triangle ADC = 37.70 = 38.

29. (B)

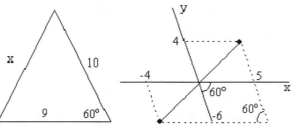

By the cosine rule:
$$x^2 = 9^2 + 10^2 - 2.9.10.\cos 60° \Rightarrow x = 9.54$$

30. (D)

$$x = 0 \Rightarrow y = \mp 3A \text{ and } y = 0 \Rightarrow x = \mp 2A$$

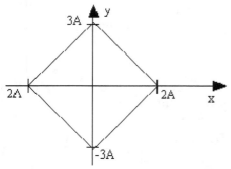

Graph represents a rhombus.

31. (B)

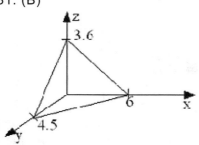

$$V = \frac{\frac{4.5 \cdot 6}{2} \cdot 3.6}{3} = 16.2$$

32. (D)

The infinite product is given by:

$$5^1.5^{\frac{1}{2}}.5^{\frac{1}{4}}.5^{\frac{1}{8}}... = 5^{1+\frac{1}{2}+\frac{1}{4}+\frac{1}{8}+....} = 5^{\frac{1}{1-\frac{1}{2}}} = 5^2 = 25$$

33. (E)

Discriminant of (9,-a,4) must be zero so that $9x^2 - axy + 4x^2$ becomes a perfect square.

$$(-a)^2 - 4 \cdot 9 \cdot 4 = 0$$

$$a^2 = 144 \Rightarrow a = \mp 12$$

34. (D)

Graph of $y = e^x - x - 2$ intersects the x axis at -1.184 and 1.146 as shown below.

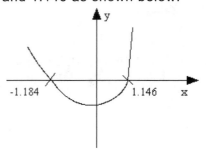

35. (C)

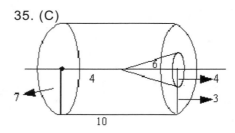

$$\pi.7^2.10 - \pi.4^2.\frac{6}{3} = 490\pi - 32\pi = 458\pi$$

36. (D)

Complex roots are 2-i and 2+i.

$$x_1 = 2 - i$$
$$x_2 = 2 + i$$
$$S = x_1 + x_2 = 4$$
$$P = x_1 \cdot x_2 = 5$$
$$\Rightarrow x^2 - Sx + P = x^2 - 4x + 5$$
$$P(x) = (x-3)(x^2-4x+5) = x^3 - 7x + 17x - 15$$

37. (D)

The inequality represents a half line whose graph is given as follows:

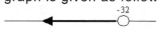

38. (D)
Graph of $y=x^3+4-3x^2$
$x^3+4-3x^2>0$ and the graph is equivalent to $-1 < x < 2$ or $x > 2$

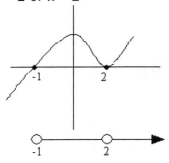

39. (A)
The required volume equals

$$\pi R^2 \cdot 2R - \frac{4}{3}\pi R^3 = 2\pi R^3 - \frac{4}{3}\pi R^3 = \frac{2\pi R^3}{3}$$

40. (E)

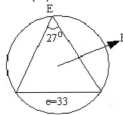

Sine rule states that
$$\frac{e}{\sin E} = \frac{33}{\sin 27°} = 2R \text{ where}$$
R is the radius of the circumscribed circle. R is approximately 72.7 for all triangles EBR. Therefore they have the some circumscribed circle.

41. (B)
$$\lim_{h \to 0} \frac{x^2 + 4x + 4 - h^2 - 4h - 4}{x-h} = \lim_{h \to 0} \frac{x^2 - h^2 + 4x - 4h}{x-h}$$
$$= \lim_{h \to 0} \frac{(x+h)(x-h) + 4(x-h)}{x-h}$$
$$= \lim_{x \to 0} \frac{(x-h)(x+h+4)}{x-h} = x + 4$$

42. (C)
$$f(x) = \frac{x^2 - x - 2}{-2x^2 + 2x + 4} = \frac{x^2 - x - 2}{-2(x^2 - x - 2)}$$
$$= \frac{(x-2)(x+1)}{-2(x-2)(x+1)} \Rightarrow f(x) = \frac{-1}{2} \text{ if and only if } x-2 \text{ and }$$
x+1 can be canceled $\Rightarrow x \neq 2$ and $x \neq -1$.

43. (B)
$2 - \dfrac{1}{x} = x \Rightarrow 2x - 1 = x^2 \Rightarrow x^2 - 2x + 1 = 0 \Rightarrow (x-1)^2 = 0 \Rightarrow x = 1$

44. (B)
$a = 5k+1$ where $k > 12$. $f(a) = (5k+1)^2 + 5k + 1 + 1$
$= 25k^2 + 10k + 1 + 5k + 2 = 25k^2 + 15k + 3$. $25k^3 + 15k$ is always even; therefore f(a) is always odd.

45. (B)
The minor and major sector cones are as follows:

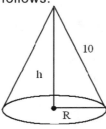

Minor sector cone:
$$2\pi R = \frac{2\pi \cdot 10}{360°} \cdot 120° \Rightarrow R = \frac{10}{3}$$
$$\Rightarrow h = \sqrt{100 - \frac{100}{9}} = 9.428$$
$$V_1 = \frac{\pi \left(\frac{10}{3}\right)^2 \cdot 9.428}{3} = 109.7$$

Major sector cone:
$$2\pi R = \frac{2\pi \cdot 10}{360°} \cdot 240° \Rightarrow R = \frac{20}{3}$$
$$\Rightarrow h = \sqrt{100 - \frac{400}{9}} = 7.454$$
$$V_2 = \frac{\pi \cdot \left(\frac{20}{3}\right)^2 \cdot 7.454}{3} = 346.925$$
$$\Rightarrow \frac{V_1}{V_2} = \frac{109.7}{346.925} = 0.3162 = \frac{\sqrt{10}}{10}$$

46. (A)
$\ln(\cos(2x)) = \ln(\cos^2 x - \sin^2 x)$
$= \ln(\cos x - \sin x)\cdot(\cos x + \sin x)$
$= \ln(\cos x + \sin x) + \ln(\cos x - \sin x)$

47. (D)
I and II satisfy the conditions:
I. $f\left(\dfrac{1}{x}\right) = \dfrac{1}{x} + \dfrac{1}{\frac{1}{x}} = \dfrac{1}{x} + x = f(x)$

II. $f\left(\dfrac{1}{x}\right) = \dfrac{1}{x} - \dfrac{1}{\frac{1}{x}} = \dfrac{1}{x} - x = -\left(x - \dfrac{1}{x}\right) = -f(x)$

$$\Rightarrow |f(x)| = \left|f\left(\frac{1}{x}\right)\right|$$

48. (B)
$$-\left(1 + \sqrt{3}\right) + \frac{1}{1 + \sqrt{3}} = \frac{\sqrt{3} - 1}{2} - 1 - \sqrt{3} = \frac{-3 - \sqrt{3}}{2}$$

49. (D)
|f(x)| reflects the negative y portion of f(x) with respect to the x axis so that the resulting graph is nonnegative throughout. We also require that |f(x)| be periodic and f(x) is not. I and III satisfy the required conditions. For the function in II, both f(x) and |f(x)| are periodic.
Answer is D.

50. (C)
f(x) has a zero at x=1; two vertical asymptotes as x= -1 and x= -2 and one horizontal asymptote at y=0.

Model Test 6 – Solutions

1. (D)
$3m = 12.26$
$m = 4.09 \Rightarrow 2m \cong 8.18$

2. (D)
Graph the function $y = e^x - \ln x$ and find its x intercept which is x = 1.48.

3. (A)
$2^0 = 3^0$
$1 = 1$
a-2b+4=0 and b-3=0
b=3 and a=2
a-b=2-3=-1

4. (B)
$x^2 - 2x + 1 = (x - 1)^2$
$x^2 + 2x + 1 = (x + 1)^2$
$(x - 1)^2 \cdot (x + 1)^2 = \left[x^2 - 1\right]^2 = \left[1 - x^2\right]^2$

5. (D)
x=0 implies that 0+4y+24=0
4y=-24 and y=-6
(0,-6) is the y intercept

6. (C)
$\dfrac{1}{y^3} = 3$
$\left(\dfrac{1}{y^3}\right)^3 = \dfrac{1}{y^9} = 3^3 = 27$

7. (E)
$x-y+2=0 \Rightarrow x-y=-2 \Rightarrow y-x=2$
$(2^{y-x})(x^2 - 2xy + 1 + y^2)$
$= (2^{y-x})(x^2 - 2xy + y^2 + 1)$
$= (2^2)(2^2 + 1) = 20$

8. (A)
$2^9 = 2^{y-1}$
$9 = y - 1$
$y = 10$

9. (E)
If $|2 - x| = x - 2$ then $2 - x$ is negative or less than zero which implies that x > 2.

10. (B)
$5xy + 2y = 3x - 10$
$2y + 10 = 3x - 5xy$
$2y + 10 = x(3 - 5y)$
$x = \dfrac{2y + 10}{3 - 5y} = \dfrac{2y + 10}{-5y + 3}$

11. (C)
If y=|x+12|+|3x+6|-12<0 is graphed, the x values for which y is negative will be observed as -5<x<1

12. (D)
$x - 2 = k \cdot (y + 2)$
$6 - 2 = k \cdot (4 + 2) \Rightarrow 4 = k \cdot 6 \Rightarrow k = \dfrac{4}{6} = \dfrac{2}{3}$
$x - 2 = \dfrac{2}{3} \cdot (y + 2)$
$x - 2 = \dfrac{2}{3} \cdot (7 + 2) \Rightarrow x = 8$

13. (C)
$a + bi = c + di \Rightarrow a = c \wedge b = d$
$(x + 3)i - (y - 4) = -5i + 6i^2 = -5i - 6$
$x + 3 = -5 \Rightarrow x = -8$
$y - 4 = 6 \Rightarrow y = 10$
10+8=2

14. (C)
Try each of the points. The point in C does not satisfy the given relation.

15. (A)
$f(g(x)) = e^{g(x)} = e^{-2x-1} \Rightarrow g(x) = -2x - 1$

16. (D)
$\dfrac{3}{5} = \dfrac{e - 1}{6} \Rightarrow 18 = 5e - 5 \Rightarrow e = \dfrac{23}{5}$

17. (C)
The point given in C is the only possibility for the remaining vertex of the parallelogram.

18. (D)
g(g(3))=g(4)=1
g(3)=4 therefore $g^{-1}(4)=3$
I and II are correct but III is false.

19. (B)
$\dfrac{\dfrac{x+1}{x}}{\dfrac{1-x^2}{x^2}} \cdot \left(\dfrac{x-1}{x}\right) = \dfrac{x+1}{x} \cdot \dfrac{x^2}{1-x^2} \cdot \dfrac{x-1}{x}$
$= \dfrac{(x+1) \cdot x^2 \cdot (x-1)}{x^2 \cdot (1-x) \cdot (1+x)} = -1$

20. (D)
There are 9·6·1=54 palindromic times in the form A:BA between 12:00 noon and 12:00 midnight. In addition 10:01, 11:11, and 12.21 are also palindromic times. So there are totally 57 palindromic times between 12:00 noon to 12:00 midnight. Therefore there are totally 57·2 = 114 palindromic times.

21. (C)
$1+x \neq 0 \Rightarrow x \neq 1$

$$\dfrac{\dfrac{1}{1+x+1}}{1+x} = \dfrac{1+x}{2+x}$$

$2+x \neq 0 \Rightarrow x \neq 2$
$x \notin \{-1, 2\}$

22. (E)
If the roots are complex numbers these roots are conjugates each other.
$(3+i, 3-i)$
$x_1 + x_2 = 3 + i + 3 - i = 6$
$x_1 \cdot x_2 = (3+i)(3-i) = 9 - i^2 = 10$
$x^2 - (x_1 + x_2)x + x_1 \cdot x_2 = x^2 - 6x + 10$

23. (A)
$$\dfrac{1-\cos^2 x}{1-\cos x} + \dfrac{1-\cos^2 x}{1+\cos x} =$$
$$= \dfrac{(1-\cos x)\cdot(1+\cos x)}{(1-\cos x)} + \dfrac{(1-\cos x)\cdot(1+\cos x)}{(1+\cos x)}$$
$$= 1 + \cos x + 1 - \cos x = 2$$

24. (C)
$h(3) = h(-2) = 8$
$g(h(3) - 5) = g(8-5) = g(3) = 4$

25. (B)
$a + ar + ar^2 + ar^3 + ... = \dfrac{a}{1-r}$ provided that $-1<r<1$

$$\text{Sum} = \dfrac{a}{1-r} = \dfrac{\dfrac{2}{3}}{1-\left(\dfrac{-2}{3}\right)} = \dfrac{\dfrac{2}{3}}{\dfrac{5}{3}} = \dfrac{2}{5}$$

26. (B)
$y^2 + y + 3x^2 + 2x = 0$ represents an ellipse since the coefficients of x^2 and y^2 are different but they have the same sign.

27. (D)
$b^2 - 4ac < 0 \Rightarrow p^2 - 4 \cdot 3 \cdot < 0 \Rightarrow p^2 < 24$
$-2\sqrt{6} < p < 2\sqrt{6}$

28. (A)
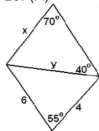
by the Cosine Rule
$y^2 = 4^2 + 6^2 - 2 \cdot 6 \cdot 4 \cdot \cos 55°$
$y^2 = 24.468 \Rightarrow y = 4.95$
by the Sine Rule
$\dfrac{4.95}{\sin 70°} = \dfrac{x}{\sin 40°}$
$x = 3.38$

29. (A)
f(x-2) shifts f(x) 2 units right.
f(x-2) + 1 shifts f(x-2) 1 unit up.

30. (C)
$$\dfrac{\pi r^2 h}{3} \cdot \dfrac{80}{100} = \dfrac{\pi \cdot (1.1r)^2 h'}{3}$$
0.8h=1.21h'
h'=0.6612h which is 33.88% less than the original height.

31. (C)
$a_n = a_1 \cdot r^{n-1}$
$a_7 = 45,000 \cdot (1-0.05)^6 = 33,079$

32. (A)
If f(x) and -f(x) intersect at no points; so f(x) does not intersect the x axis which implies that the discriminant is negative:
$b^2 - 4ac < 0$ and $b^2 < 4ac$.

33. (B)
The matrix in choice B satisfies both equations.

34. (C)
$$f(f(x)) = \dfrac{a \cdot \left(\dfrac{ax-b}{c}\right) - b}{c} = \dfrac{a^2 x - ab - cb}{c^2} = x$$
$\dfrac{a^2}{c^2}x - \dfrac{ab+cb}{c^2} = x$ and $\dfrac{a^2}{c^2} = 1$; so $a^2 = c^2$ which implies that |a| = |c|. Moreover a must be nonzero because if a is zero then f(f(x)) will be a constant function.

35. (D)
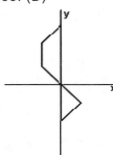
Correct answer is given in D and will be obtained by rotating the above figure for a 90° counter clockwise angle.

36. (B)
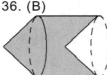
The left hand cone fills the conical gap in the right hand making a complete cylinder.
$\pi r^2 h = \pi \cdot (3\sqrt{3})^2 \cdot 10 = 270\pi$

37. (E)
$a - 5 = 9^x \Rightarrow a - 5 = 3^{2x}$
$3^x = 3 - b$
$(3 - b)^2 = a - 5$
$9 - 6b + b^2 = a - 5$
$a = 14 - 6b + b^2$

38. (D)
Correct answer is given in D.

39. (B)
If b = 1 then $\frac{a}{b} = a$.
$2 \cdot 5^a = 3 \cdot 4 = 12$.
If the function $y = 2 \cdot 5^x - 12$ is graphed, its x intercept will be observed at x=1.11which is the correct answer.

40. (D)
$f(x) = mx + n$
$2m + n = 5$
$5m + n = 2$
Therefore m = -1 and n = 7.
f(x) = -x + 7 and f(3) = -3 + 7 = 4.

41. (B)
Between 4:00 PM and 6:00 PM the volume V_B of B added to the mixture is not a linear function of time; density of the mixture is a linear function of time.

42. (E)
All of the figures can be obtained.

43. (C)
$x = 1 + t$ $t = x - 1$ $x - 1 = y - 3$
$y = 3 + t$ $t = y - 3$ $x + 2 = y$
(1, 3) and (2, 4) satisfy x+2=y

44. (D)
Range consists of {0,1,2} which is the set of non negative integers less than 3.

45. (B)
$f^{-1}$of is the identity function provided that f(x) is an invertible function. f(x) is invertible since when it is graphed, it passes the horizontal line test. Therefore $f^{-1}$of=I and $(f^{-1}$of$)(x^2) = x^2$

46. (D)
I and II are correct; III is not correct because both sets have the same standard deviation.

47. (B)
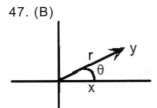
$r \cdot \cos\theta = x; r \cdot \sin\theta = y;$
and $r^2 = x^2 + y^2$
$r = \cos\theta + \sin\theta$
$r^2 = r \cdot \cos\theta + r \cdot \sin\theta$
$x^2 + y^2 = x + y$

48. (D)
$\binom{4}{2} \cdot 3 + \binom{6}{2} = 33$

49. (C)
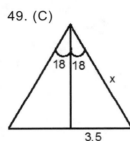
$\frac{3.5}{x} = \sin 18° \Rightarrow x = 11.33$
perimeter $= 2 \cdot 11.33 + 7 = 29.7$

50. (C)
$f(x) = \dfrac{1}{x^2 - 1}$
$x^2 - 1 = 0 \Rightarrow x = \pm 1$
f(x) has 2 vertical asymptotes
$\lim_{x \to \infty} \dfrac{1}{x^2 - 1} = 0$
y=0 is the horizontal asymptote of f(x).

Model Test 7 – Solutions

1. (B)
$1 + \dfrac{1}{x} = 2 + \dfrac{2}{x} \Rightarrow 1 - 2 = \dfrac{2}{x} - \dfrac{1}{x}$
$-1 = \dfrac{1}{x} \Rightarrow x = -1$

2. (E)
$\left(\dfrac{1}{b} + \dfrac{1}{c} \right) : \dfrac{1}{a} = \dfrac{b + c}{bc} \cdot a = \dfrac{ab + ac}{bc}$

3. (C)
A few periods of the function y=2cos(3x) can be graphed and the coordinates of P will be observed as (2π/3, 2).

4. (A)

Z

3 miles

X

4 miles

Y

5. (E)
$$2x + \frac{3}{3(2-x)} = 4 - \frac{5}{5(x-2)}$$
$$\Rightarrow 2x + \frac{1}{2-x} - \frac{1}{2-x} = 4 \Rightarrow x = 2$$
However $x = 2$ makes the denominator zero therefore solution set is empty.

6. (E)
$$g(f(3)) = g\left(\sqrt{3^2 \cdot (3+1)}\right) = g(6) = 6^2(6+1) = 252$$

7. (D)
$$r\sin\theta = y \Rightarrow r = \frac{y}{\sin\theta}$$

8. (C)
$$5ab^2c^3 = \frac{a^2b^2}{c^{-3}} \Rightarrow \frac{5b^2c^3c^{-3}}{b^2} = \frac{a^2}{a} \Rightarrow 5 = a$$

9. (E)
Since the ordinate of the point B is not known, the slope of segment BO cannot be determined from the information given.

10. (C)
$$\underbrace{\sin^2 x + \cos^2 x}_{1} + 2\sin x \cdot \cos x$$
$$+ \underbrace{\sin^2 x + \cos^2 x}_{1} - 2\sin x \cdot \cos x = 2$$

11. (D)
$$\left.\begin{array}{l} x_1 = 5 \\ x_2 = -3 \end{array}\right\} (x-5)(x+3) = x^2 - 2x - 15 \text{ is a factor of}$$
$P(x)$.

12 (B)
100a+b-x=100b+a
x=99(a-b)
Therefore a-b can be at most 8 when a=9 and b=1; and x=99·8=792 cents.

13. (B)
$$2g(10) - 5 = 5 \Rightarrow 2g(10) = 10 \Rightarrow g(10) = 5.$$

Therefore g(x) can equal x-5.

14. (B)

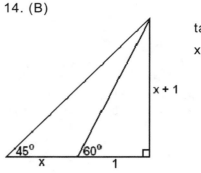

$\tan 60° = x+1 = \sqrt{3}$
$x = \sqrt{3} - 1$

x + 1

45°   60°

x   1

15. (D)
$$AD = \sqrt{2^2 + 4^2} = 2\sqrt{5}$$
$$\text{Area of ABCD} = 4 \cdot 2\sqrt{5} = 8\sqrt{5} = 17.89$$

16. (D)
sinx=0.613 implies that x=37.8° therefore cos(37.8°/2)=0.946

17. (C)
Equation of the line m is x = -3.

18. (B)
The password must contain one even and two odd integers. One of the odd numbers must be a multiple of 3 and the other one must be less than or equal to 12. The only possibility is 11-27-14.

19. (E)
$$r = 4 \Rightarrow d = 8 = a\sqrt{3} \Rightarrow a = \frac{8}{\sqrt{3}}$$

surface area equals
$$6 \cdot \left(\frac{8}{\sqrt{3}}\right)^2 = 6 \cdot \frac{64}{3} = 128 \text{ square inches.}$$

20. (C)
$$g(f(x)) = 2(2x^2 - 3x) + 5 = 2 \cdot f(x) + 5$$
So $g(x) = 2x + 5$

21. (C)
$$(5-x)(50-x) = (15-x)^2$$
$$250 - 55x + x^2 = 225 - 30x + x^2$$
$$25 = 25x \Rightarrow x = 1$$

22. (A)
$$f(10) = 0 \cdot x^2 + 0 \cdot x + c = 4 \Rightarrow c = 4$$
$$f(1) = a + b + c = 3 \Rightarrow a + b = 3 - c = 3 - 4 = -1$$

23. (C)
By the cosine rule

$\cos x = \dfrac{48^2 + 56^2 - 40^2}{2.48.56} \cong 0.714$

$\Rightarrow \theta = \arccos(0.714) \Rightarrow \theta \cong 44.42°$

24. (D)

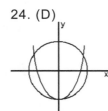

The given curves intersect at three points as indicated in the figure.

25. (E)
$16 - x^2 > 0 \Rightarrow x^2 < 16 \Rightarrow |x| < 4 \Rightarrow -4 < x < 4$

26. (C)

$\sin\theta = \dfrac{1}{\tan\theta} \Rightarrow \sin\theta \cdot \dfrac{\sin\theta}{\cos\theta} = 1$

$\Rightarrow \sin^2\theta = \cos\theta \Rightarrow 1 - \cos^2\theta = \cos\theta$

$\Rightarrow \cos^2\theta + \cos\theta - 1 = 0$

So $\cos\theta = \dfrac{-1+\sqrt{5}}{2} \Rightarrow \theta = 0.9045 \approx 0.905$

27. (C)
$\log_2 2 = 1$ and $\log_2 4 = 2$ and $\log_2 6 = 2.58$
Area $= 2 + 2 \cdot 2 + 2 \cdot 2.58 = 11.17$}

28. (B)
$x + 1 = 0 \Rightarrow x = -1$ is the vertical asymptote.

$\displaystyle\lim_{x \to} \dfrac{2-x}{x+1} = -1$ Therefore $y = -1$ is the horizontal asymptote.

29. (C)
$f(x + 1 - 1) = (x+1)^2 - 1 \Rightarrow f(x) = x^2 + 2x + 1 - 1 = x^2 + 2x$

30. (C)
If (0, 0) is a point on the circle then it passes through the origin and it can be either tangent to one of the axes or not tangent to any of them at all.

31. (D)

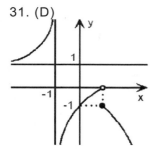

Range of $f(x): R - [0,1]$

32. (B)
$2x - 3y + 5 = 0 \Rightarrow \dfrac{2x+5}{3} = y = 2x^2 \Rightarrow 6x^2 - 2x - 5 = 0$

By the quadratic formula $x = 1.09463$ or $x = -0.761294$

33. (D)
$y = \log_3 x \Rightarrow x = \log_3 y \Rightarrow y = f^{-1}(x) = 3^x$

34. (C)
$x_1 = 1$

$x_2 = \sqrt{2 \cdot 1 + 3} = 2.236$

$x_3 = \sqrt{2 \cdot 2.236 + 3} = 2.733$

$x_4 = 2.909$

$x_5 = 2.969$

so k=4 is the greatest possible value of k.

35. (E)
Area of rectangle
$= r\sin\theta \cdot 2r\cos\theta = r^2 \cdot 2\sin\theta \cdot \cos\theta = r^2 \cdot \sin(2\theta)$

36. (D)
$1 - 0.3^3 = 0.973$

37. (A)
the magnitude of vector **a** − **b** cannot exceed 15 + 8 and must not be less than 15 − 8.

38. (C)
$4.14^a = 2.07^b$
$\log 4.14^a = \log 2.07^b$
$a \cdot \log 4.14 = b \cdot \log 2.07$
$\dfrac{a}{b} = \dfrac{\log 2.07}{\log 4.14} = 0.512$

39. (C)
A new plane is introduced in such a way that it doesn't contain the line. This now plane introduces n more regions.

40. (D)
Answer is given correctly in D.

41. (A)
$\sin 2x = \dfrac{1}{2} \Rightarrow 2x = \dfrac{\pi}{6}$ or $2x = \dfrac{5\pi}{6} \Rightarrow x = \dfrac{\pi}{12}$ or $x = \dfrac{5\pi}{12}$

42. (C)
$\sin\theta + \sin(90° - \theta) + \sin(-\theta) + \sin(90° + \theta)$
$= \sin\theta + \cos\theta - \sin\theta + \cos\theta = 2\cos\theta$

43. (D)
$\dfrac{(n+r+1)!}{(n+r-1)!} = \dfrac{(n+r+1) \cdot (n+r) \cdot (n+r-1)!}{(n+r-1)!} = (n+r)^2 + n + r$

44. (E)

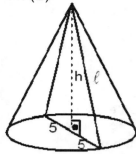

$$\ell = \frac{36 - 2 \cdot 5}{2} = 13$$

$$h = \sqrt{13^2 - 5^2} = 12$$

45. (C)
For a conditional proposition p⇒q, the converse is q⇒q

46. (E)
When f(x) translated 2 units up, we get f(x)+2 and when f(x)+2 is translated 2 units left, we get up f(x+2)+2.
$$g(x) = f(x+2) + 2 \Rightarrow g(-2.1) = f(-2.1+2) + 2 =$$
$$f(-0.1) + 2 = -(-0,1)^2 + 3 + 2 = 4.99$$

47. (A)

1st basket	2nd basket
0	9
1	8
2	7
3	6
4	5

Since the baskets are identical there are five possibilities.

48. (D)
I. The set of negative real numbers does not have a greatest element.
II. 97
III. $-100 \le x - 1 \le 100 \Rightarrow -99 \le x \le 101$. So, the greatest element is 101.

49. (C)
$$\frac{x^2}{50} + \frac{y^2}{20} = \frac{1}{10} \Rightarrow \frac{x^2}{5} + \frac{y^2}{2} = 1 \Rightarrow b^2 = 2 \Rightarrow b = \sqrt{2}$$
$$2b = 2\sqrt{2} \approx 2.828$$

50. (C)

There are two ways get $\frac{a+b}{ab}$ negative.
i. a<b<0 or ii. a<0<b and |a|<|b|.

Model Test 8 – Solutions

1. (E)
$\frac{a}{b}$ equals zero if and only if a is zero and b is nonzero. Therefore the given equation has no solution.

2. (D)
$$\log_2 9 = \frac{\log 9}{\log 2} = 3.169 \approx 3.17$$

3. (B)
$$(-243)^{\frac{-1}{5}} = \frac{-1}{3}$$

4. (E)
$$\frac{x^{(-2)(-5)}}{x^{2+5}} = \frac{x^{10}}{x^7} = x^{10-7} = x^3$$

5. (B)
$$|3x - 1| \ge 7 \Rightarrow 3x - 1 \ge 7 \text{ or } 3x - 1 \le -7 \Rightarrow x \ge \frac{8}{3} \text{ or } x \le -2$$

6. (B)
$$\sqrt{p^3} = 10 \Rightarrow p^3 = 10^2 \Rightarrow p = \sqrt[3]{100} \cong 4.64$$

7. (B)
$$\frac{\frac{2x - 1 - 2x}{2x}}{\frac{4x^2 - 4x^2 + 1}{2x}} = \frac{-1}{1} = -1$$

8. (B)
If $y = \frac{x+1}{(x-3)(x+3)}$ is graphed then the set of x values for which y is positive will be observed as -3 < x < -1 or x > 3.

9. (D)
$$y^2 + 4y + 4 - 4 = x^2 + 2x + 1 - 1 \Rightarrow (y+2)^2 - 4 = (x+1)^2 - 1$$
$$\Rightarrow (y+2)^2 - (x+1)^2 = 3 \Rightarrow \frac{(y+2)^2}{3} - \frac{(x+1)^2}{3} = 1$$
The above equation represents a hyperbola.

10. (C)

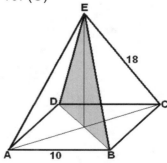

$$|DB| = |AC| = \sqrt{10^2 + 10^2}$$
$$= 10\sqrt{2}$$
$$\Rightarrow \frac{1}{2}|DB| = 5\sqrt{2}$$
$$|EC| = |EB| = 18$$

Let F be the midpoint of DB. So,
$$\sin(F\hat{E}B) = \frac{5\sqrt{2}}{18} \Rightarrow m(F\hat{E}B) \approx 23.13°$$
$$m(D\hat{E}B) = 2 \cdot m(F\hat{E}B) = 2 \cdot 23.13° \approx 46.26°$$

11. (D)
The are two steps to draw f(|x|): First, omit the negative x portion of the graph; then reflect the positive x portion of the graph with respect to the y-axis.

12. (D)
g(x, y) = g(-x, y) implies that g(x, y) is symmetric with respect to the y axis.
g(x, y) = g(-x, -y) implies that g(x, y) is symmetric with respect to the origin.

13. (A)
$y = 3 \cdot \sin^2(2x) + 1$ can be graphed in the radian mode and the x-distance between two adjacent maxima will give the period which is $\dfrac{\pi}{2}$.

14. (B)
$$x^2 + 4x + 4 - 4 + y^2 + 2y + 1 - 1 = 0$$
$$(x+2)^2 + (y+1)^2 = 5$$
The equation above represents a circle whose center is at (-2,-1) and whose radius has a length of $\sqrt{5}$.

15. (B)
For x<0, the function is a decreasing function and for x>0 the function is an increasing function. So, for x<0, the graph of the slopes of tangents must be negative (below x-axis), and for x>0, the graph of the slopes of tangents must be positive (above x-axis). Since the function is quadratic, the graph of the slopes of tangents must be linear.

16. (E)
For
$$f(x) = e^{-x} \cdot \left(2x^3 - 4x + 1\right) < 1 \Rightarrow e^{-x}\left(2x^3 - 4x + 1\right) - 1 < 0$$
The graph of $y = e^{-x} \cdot \left(2x^3 - 4x + 1\right) - 1$ looks like the following and the values of x for which y is negative are x<1.5 or 0<x<1.86 or x>6.03.

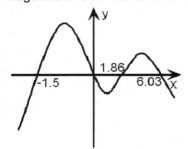

17. (E)

11 AM	1 PM
12 AM	2 PM
1 PM	3 PM
2 PM	4 PM
3 PM	5 PM
4 PM	6 PM

18. (A)
$$|2x - 6| - 2 \neq 0 \Rightarrow |x - 3| \neq 1 \Rightarrow x - 3 \neq 1 \Rightarrow x \neq 4$$
$$\Rightarrow x - 3 \neq -1 \Rightarrow x \neq 2$$
So the domain is $R - \{2,4\}$

19. (B)
Radius is $\dfrac{15 - (-3)}{2} = 9$ and center is at
$\dfrac{15 + (-3)}{2} = 6$. $|x - 6| > 9$ since the points
-3 and 15 are not included in the solution set.

20. (C)
If we replace (x, y) by (-x,-y) we will get the reflection of the curve about the origin.
$$-y = (-x)^3 - 2(-x) + 1 \Rightarrow -y = -x^3 + 2x + 1$$
$$\Rightarrow y = x^3 - 2x - 1$$

21. (C)
$$(x^2 - x)(-2x - 1) - (x^2 + x)(2x + 1) = 0$$
$$\left(x^2 - x + x^2 + x\right)\left(-2x - 1\right) = 0$$
$$2x^2(2x + 1) = 0$$
$$2x^2 = 0 \Rightarrow x = 0$$
$$2x + 1 = 0 \Rightarrow x = \dfrac{-1}{2}$$

22. (C)
$$|VS| = 20 \cdot \tan 22° = 8.08$$
$$|FV| = \dfrac{20 - 5.5}{\tan 14°} = 58.15$$
$$|VS| + |VF| = 8.08 + 58.15 = 66.23\,\text{ft}$$
$$66.23\,\text{ft} = 66\,\text{ft} + 0.23\,\text{ft}$$
$$= 66\,\text{ft} + 0.23 \cdot 12\,\text{inches}$$
$$= 66\,\text{ft} + 2.84\,\text{inches}$$

23. (A)
$$\sec^2 x - 1 = \tan^2 x$$

24. (B)
$$a_6 = -144$$
$$r = -2$$
$$a_1 \cdot r^5 = -144$$
$$a_1 = -144 / (-2)^5 = 4.5$$
$$a_2 = a_1 \cdot r = 4.5 \cdot -2 = -9$$

25. (A)
The given equations are both satisfied when tanx and sinx have the same signs; which is in the first and fourth quadrants.

26. (D)
$P(-2) = 12.$

$x = -1$

$P(-2) = P(-1-1) = (-1)^3 - 2 \cdot (-1)^2 + 3h \cdot (-1) = 12$

$-1 - 2 - 3h = 12 \Rightarrow -3h = 15 \Rightarrow h = -5$

27. (D)
$\binom{10}{2} = \dfrac{10 \cdot 9}{2} = 45$

28. (C)
$x - 3 \geq 0 \Rightarrow x \geq 3$; $\sqrt{5}$ does not meet this restriction.

29. (C)
Since B is a subset of A we can omit set B. On the other hand each element of A is different.

So, there are $\binom{8}{2} = \dfrac{8 \cdot 7}{2} = 28$ different sums.

30. (C)
Let x be the distance in miles between *RUSH* Academy and Mathville; let V_B and V_C be the velocities of the student busses B and C, respectively.

So $\dfrac{x-80}{V_B} = \dfrac{x-144}{V_C}$ where $V_B = 5V$ and $V_C = 4V$.

Then, $\dfrac{x-80}{5V} = \dfrac{x-144}{4V}$ we get x=400 miles.

31. (D)

1		2		6		15		31		56		**92**
	1		4		9		16		25		36	

The difference between two successive terms of this sequence is always a perfect square. Therefore the seventh term of the sequence will be 92.

32. (B)
Equation of the parabola can be written as $x = a \cdot (y-2)^2$. (12, 0) is a point on the parabola so $12 = a \cdot (0-2)^2 \Rightarrow a = 3$. Equation of the parabola is $x = 3(y-2)^2$.

33. (E)

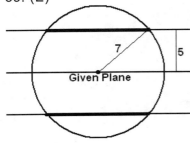

Given Plane

When a sphere is intersected by a plane, the intersection is a circle. In this case the sphere is intersected by two planes and the intersection is two circles.

34. (A)
The points (0, 2) and (1, 0) must satisfy the function $y = x^2 + mx + n$. Therefore $m+n = -1$ and $n = 2$ which implies that m=-3. As a result the function is $y = x^2 - 3x + 2$. Coordinates of the vertex can be determined by a graphical calculator by finding the local minimum point which is $(1.5, -0.25)$.

35. (C)
It can be observed that $f(x) = f(x+5)$ and that the period of the function is 5. So,
$$\left.\begin{array}{l} f(140) = f(0) = 1 \\ f(142) = f(2) = 2 \end{array}\right\} f(140) + f(142) = 1 + 2 = 3$$

36. (D)
I. $\sqrt{y} = \sqrt{4x^2} \Rightarrow |y| = |2x| \Rightarrow y = \pm 2x$

II. $\sqrt[3]{y^3} = \sqrt[3]{(2x)^3} \Rightarrow y = 2x$

III. $\left(\sqrt{y}\right)^2 = \left(\sqrt{2x}\right)^2 \Rightarrow y = 2x$

37. (E)
Correct answer is given in E.

38. (E)
$x^2 - 1 = 0 \Rightarrow (x-1)(x+1) = 0 \Rightarrow x = \pm 1$ are the vertical asymptotes. The limit at infinity is 1 so y = 1 is the horizontal asymptote.

39. (A)
The center of circle is (1,-1). Another vertex may be symmetric with the given one with respect to the center of the circle. Since the midpoint of the segment whose endpoints are (3, -3) and (-1, 1) is the center of the circle, then (-1, 1) is another vertex of the square.

40. (C)
$y = \dfrac{2x+7}{x+4}$ can be rewritten as $y = 2 - \dfrac{1}{x+4}$ which implies that $y - 2 = \dfrac{-1}{x+4}$ and that (-4, 2) is the center of the symmetry.

41. (A)
$x = e^{2y} + 1 \Rightarrow x - 1 = e^{2y}$

$\Rightarrow \ln(x-1) = 2y \Rightarrow f^{-1}(x) = \dfrac{1}{2}\ln(x-1)$

42. (B)
In the worst case he will withdraw eight socks each having a different color and the next one that he withdraws will match one of the socks he already has.

43. (A)

Among 36 possible sums, we should exclude the pairs (1,5), (1,4), (1,3), (1,2), (2,4), (2,3) each of whose sum is less than or equal to six.

Therefore the answer is $\dfrac{36-6}{\binom{9}{2}} = \dfrac{30}{\frac{9\cdot8}{2}} = \dfrac{30}{36} = \dfrac{5}{6}$.

44. (B)

$$\left.\begin{array}{l}\dfrac{f(x)+1}{2}=4^x \\[2mm] g(x)-1=4^x\end{array}\right\}\quad g(x)=\dfrac{f(x)+3}{2}$$

45. (B)

$2x - 1 = A$

$3x + 4 = B$

If we eliminate x we get $B = 1.5A + 5.5$

46. (B)

$\sin 2x \cdot \cos 2x = \dfrac{1}{2}\cdot 2\cdot \sin(2x)\cos(2x) = \dfrac{1}{2}\cdot \sin(4x)$

Therefore the maximum value of $\sin(2x)\cdot\cos(2x)$ is 0.5.

47. (A)

$\tan A \cong \dfrac{1}{4}\cdot 5.67 = 1.418 \Rightarrow A = \tan^{-1}(1.418) = 0.957$

Please note that correct answer is in radians.

48. (C)

$\sin^2 x < \cos x \Rightarrow \sin^2 x - \cos x < 0$ If we sketch $y = \sin^2 x - \cos x$ with the graphing calculator we will get a graph like the following.

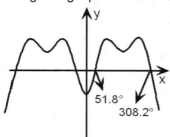

The values of x for which f(x) is negative are $0° \le x < 51.8°$ or $308.2° < x \le 360°$.

49. (E)

$A(a,b)\quad B(a,-b)\quad C(b,-a)\quad .D(-b,-a)$

$|AD| = \sqrt{(a-(-b))^2 + (b-(-a))^2} = \sqrt{2\cdot(a+b)^2}$

50. (B)

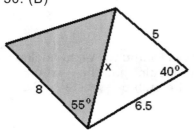

By the cosine rule

$x^2 = 5 + (6.5)^2 - 2\cdot 5\cdot(6.5)\cdot\cos 40° \Rightarrow x = 4.17$

$\text{Area} = \dfrac{1}{2}\cdot x\cdot 8\cdot\sin 55° = 13.69$

Model Test 9 – Solutions

1. (E)

$y + x = 0 \Rightarrow y = -x$

$x\cdot(-x) + 1 = 0 \Rightarrow x^2 = 1 \Rightarrow \begin{cases}x=1\\y=-1\end{cases}$ or $\begin{cases}x=-1\\y=1\end{cases}$

2. (D)

$-1 < k < 0$

$1 > -k \Rightarrow \dfrac{-1}{k} > 1$

3. (C)

$A = 4k+2$

$A^2 = 16k^2 + 16k + 4$

$= 8(2k^2+2k)+4$

$= 8P + 4 \Rightarrow$ remainder is 4.

4. (B)

$-3+k+4 = -3-k+4$

$2k = 0$ and $k = 0$

5. (C)

$f(g(h(5))) = f(g(15)) = f\left(\dfrac{15-1}{2}\right) = f(7) = 2.7 + 1 = 15$

6. (E)

$\dfrac{240}{\frac{120}{60}+\frac{120}{V}} = 72 \Rightarrow \dfrac{10}{3} = 2 + \dfrac{120}{V} \Rightarrow \dfrac{4}{3} = \dfrac{120}{V} \Rightarrow V = \dfrac{360}{4} = 9$

7. (B)

$f(1-x) = (1-x)^2 - 1 = 1 - 2x + x^2 - 1 = x^2 - 2x = x(x-2)$

8. (D)

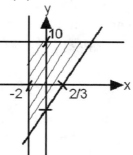

9. (D)

$m(\hat{CAB}) = y$

and

$m(\hat{ACB}) = 90° \Rightarrow x + y = 90° \Rightarrow \cos x = \sin y = \dfrac{4}{5}$

10. (D)
The lines bounding the shaded region are: 2x + 3y = 6; x = 1; x = 2; y = 0

11. (C)

If a right triangle is rotated 360° about one of its legs, the solid generated will be a cone.

12. (D)
I. $g(g(1))=g(2)=3$: correct
II. $g(g^{-1}(1))=g(4)=1$: correct
III. $g^2(1)=[g(1)]^2=2^2=4$: incorrect

13. (D)
$\dfrac{5\pi}{6}$ radius $=\dfrac{5}{6}\cdot 180°=5\cdot 30°=150°$

14. (E)

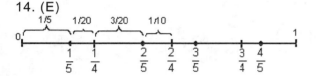

15. (B)

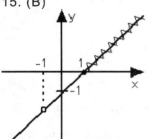

$y=\dfrac{x^2-1}{x+1}=\dfrac{(x-1)(x+1)}{x+1}$
and $x\neq -1$

16. (D)
$\dfrac{\pi ab-\pi b^2}{\pi b^2}=\dfrac{3}{4}\Rightarrow \dfrac{a}{b}=\dfrac{7}{4}$

17. (E)
$\dfrac{\sec\theta}{\sin\theta}=\dfrac{\frac{1}{\cos\theta}}{\sin\theta}=\dfrac{1}{\sin\theta\cdot\cos\theta}$

18. (A)

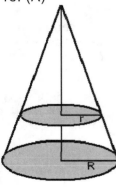

Let V_1 and V_2 be the volumes of the upper and lower segments respectively. By the rule of similarity, the ratio of the volumes of two similar solids is the cube of the ratio of their side lengths.

$\dfrac{V_1}{V_1+V_2}=\left(\dfrac{2}{3}\right)^3$ which

implies that $\dfrac{V_1}{V_2}=\dfrac{8}{21}$.

19. (B)
$C = A_{4\times 5}\cdot B_{5\times 2}=(AB)_{4\times 2}$

20. (D)

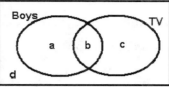

	Boys	Girls	Totals
TV	18 = b	9	27
Other than TV	7	6 = d	13
Totals	25	15	40

b+d=18+6=24

21. (E)
Graph of $y=\sin\left(\dfrac{\pi}{6}\right)-\cos\left(\dfrac{\pi}{6}+x\right)$ in the interval $0<\theta\le \dfrac{3\pi}{2}$ intersects the x axis at $\dfrac{3\pi}{2}$.

22. (E)
x = 10 makes the denominator zero therefore it must be excluded from the domain.

23. (D)
$2^x.4=k\cdot 2^{x-1}\Rightarrow 2^{x+2}=k\cdot 2^{x-1}\Rightarrow k=\dfrac{2^{x+2}}{2^{x-1}}=2^3=8$

24. (E)
When x=2 f(x) is minimum and the minimum value equals 3.

25. (B)
center is at (0,3); radius is 4.

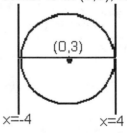

4=k-1and k=5 or -4=k-1 and k=-3

26. (D)
f(-x) = 2(-x)² = 2x²; correct graph is given in D.

27. (E)
Mean and median will be both increased by 20 but the standard deviation will be unaltered.

28. (A)
3.1³-2.1²+k.1+5=0; k+6=0 and k=-6

29. (B)
$f(-k)=f(k)\Rightarrow k^2-3k-10=k^2+3k-10\Rightarrow k=0$

30. (E)
$\theta=\dfrac{2\pi}{3}$ satisfies the given inequality. Please make sure that you do the calculation in the radian mode.

31. (C)
$x=\sec\theta\Rightarrow\dfrac{1}{x}=\cos\theta\wedge y=\csc\theta\Rightarrow\dfrac{1}{y}=\sin\theta$

$\cos^2\theta+\sin^2\theta=1$ implies that $\left(\dfrac{1}{x},\dfrac{1}{y}\right)$ are points on the unit circle.

32. (D)
$\dfrac{a^{x+1}}{a^x}=8\Rightarrow a^{x+1-x}=8\Rightarrow a=8$

33. (E)
$y=\sec(x)-1$

34. (B)
$\cos x=\dfrac{\sqrt{2}}{2}\Rightarrow x=\dfrac{\pi}{4}\Rightarrow\cos(3x)=\cos\left(\dfrac{3\pi}{4}\right)=\dfrac{-\sqrt{2}}{2}$

35. (B)
$\dfrac{n!+(n-1)!}{(n+2)!-(n+1)!}=\dfrac{(n-1)!\cdot(n+1)}{(n+1)!\cdot(n+2-1)}=\dfrac{(n-1)!}{(n+1)!n(n-1)!}=\dfrac{1}{n^2+n}$

36. (D)
$f(x)=1+\dfrac{1}{1-\dfrac{1}{x}}=1+\dfrac{x}{x-1}=\dfrac{x-1+x}{x}=\dfrac{2x-1}{x}$

Range does not include 2 because y=2 is a horizontal asymptote. Since x=1 is not in the domain then f(x) cannot equal $\dfrac{2\cdot1-1}{1}=1$

37. (C)
$\begin{array}{l}a<b\\ \underline{b<c}\\ a+b<b+c\end{array}\Rightarrow$
is correct and a counter example can be given for all others.

38. (B)
$\sum_{k=0}^{7}(k-2)=\sum_{k=0}^{7}k-\sum_{k=0}^{7}2=\sum_{k=0}^{7}k-8\cdot2=\sum_{k=0}^{7}+A\Rightarrow A=-16$

39. (A)

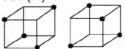

The cube can be partitioned in two unique ways to make a tetrahedron each of whose faces is an equilateral triangle. Each tetrahedron has four faces therefore $2\times4=8$ equilateral triangles can be made.

40. (E)
ABCx1001=ABCABC
The arithmetic mean of the digits of ABC is $\dfrac{A+B+C}{3}$. The arithmetic mean of the digits of ABCABC is $\dfrac{2A+2B+2C}{6}$ both of which are the same.

41. (D)
51, 15, 62, 26; therefore probability is $\dfrac{4}{36}=\dfrac{1}{9}$.

42. (C)
$\lim_{n\to\infty}\dfrac{n+\sqrt{4n^2+1}}{2n-1}=\lim_{n\to\infty}\dfrac{n+2n}{2n}=\dfrac{3}{2}$

43. (A)
N is the least common multiple of 54, 55 and 56; therefore $N=2^3\cdot3^3\cdot5\cdot7\cdot11$.

44. (D)
base$=AC=\sqrt{2}$ and area$=\dfrac{b\cdot h}{2}=\dfrac{\sqrt{3}}{2}$

$\dfrac{\sqrt{2}\cdot h}{2}=\dfrac{\sqrt{3}}{2}$ therefore $h=\dfrac{\sqrt{3}}{\sqrt{2}}=AC\cdot\dfrac{\sqrt{3}}{2}$ and triangle is equilateral. This implies that the third vertex is G.

45. (A)
i^{4n+k} can be -1 only if k given the remainder of 2 when divided by 4.

46. (D)
$\overrightarrow{OB}-\overrightarrow{OA}=(8,4)$ and the length of the resultant vector is $\sqrt{8^2+4^2}=\sqrt{80}$.

47. (D)
The distance from an arbitrarily selected point (x, y) and the x axis is equal to the distance between this point and the point (-3, 4). Therefore

$|y|=\sqrt{(x+3)^2+(y-4)^2}$

$y^2=x^2+6x+9+y^2-8y+16$

$8y-16=(x+3)^2$

$8(y-2)=(x+3)^2$

48. (B)

shaded area = $\dfrac{3}{4} \cdot \left(\dfrac{1}{2} \cdot p \cdot q \cdot \sin\theta\right) \cdot 2$ and it equals

$\dfrac{3}{4} \cdot p \cdot q \cdot \sin\theta$.

49. (D)

$b^2 - 4ac < 0$ implies that $f(x) = ax^2 + bx + c$ does not intersect the x axis.

50. (D)

The statement logically equivalent to $p \Rightarrow q$ is its contrapositive, $q' \Rightarrow p'$, which is given in C.

Model Test 10 – Solutions

1. (B)
$f(a+b) = f(a) \cdot f(b)$

$x^{a+b} = x^a \cdot x^b$

$f(x)$ must be an exponential function of the form x^a. Therefore $f(0) = x^0 = 1$.

2. (D)
$f(-3) = -3 + 3 \cdot (-3)^3 = -3 - 81 = 84$

3. (E)
$\sqrt{a-b} = c - d$ implies that $a - b \geq 0$ and $c - d \geq 0$. The statement in E need not be satisfied.

4. (A)
$3(x+y) = -9 \Rightarrow x + y = -3$
$4(y+z) = 8 \Rightarrow y + z = 2$
$x + y - (y + z) = x - z = -3 - 2 = -5$.

5. (E)

A. $p^3 = p \cdot p \cdot p$ is not prime.
B. $p \cdot q$ has p and q as factors other than 1 and $p \cdot q$ and it is not prime.
C. $7p$ has 7 and p as factors other than 1 and $7p$ and it's not prime.
D. $p + q$ is an even number greater than 2 and it is divisible by 2.
E. For $p = 5$ and $q = 3$; $p - q = 5 - 3 = 2$ and it is prime.

6. (A)
$\left(\sqrt[3]{p}\right)^2 = 7 \Rightarrow \sqrt[3]{p} = \mp\sqrt{7} \Rightarrow p = \left(\mp\sqrt{7}\right)^3 = \mp7\sqrt{7} = \mp18.52$

7. (D)
$c_A = r_B$ must hold; the number of columns of the first matrix must be equal to the number of rows of the second one.

8. (C)
$\dfrac{2 + 4 + 6 + 8 + 10 + 12 + x}{7} = 10 \Rightarrow x = 28$

9. (E)
Longer side of the rectangle is $\sqrt{5^2 + 12^2} = 13$, therefore shaded Area equals $7 \cdot 13 = 91$.

10. (B)

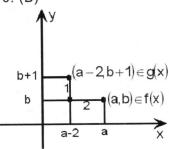

The function $f(x)$ is translated 2 units left and 1 unit up to get $g(x)$. Therefore the new point is $(a-2, b+1)$.

11. (D)

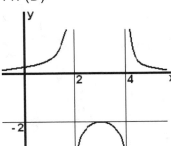

As it can be observed from the graph of $f(x)$ given above, the range of $f(x)$ is given by $y \leq -2$ or $y > 0$.

12. (B)
Let Set X contain the numbers 1, 2, 3, 4, 5, 6, 7, 8, 9, 10, 11, 12, and 13.

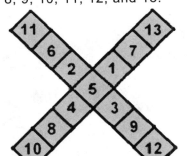

A can be any odd number in set X (please observe the figure given)

13. (C)
7 other people will be chosen among the remaining 22 people; therefore there are $\binom{22}{7}$ possible ways.

14. (D)
$y = -5x + 2 \Rightarrow 2(-5x + 2) - 3x = 4$
$\Rightarrow -10x + 4 - 3x = 4 \Rightarrow x = 0 \Rightarrow y = -5 \cdot 0 + 2 = 2$
the intersection point is (0,2)

15. (D)

The expression in III can be modified as follows:

$\sqrt[4]{y^4} = \sqrt[4]{(2x)^4} \Rightarrow |y| = |2x| \Rightarrow y = 2x$ or $y = -2x$

Therefore III is not implied.

16. (A)

If we replace (x, y) with (x, -y), we get the reflection of the curve about the x-axis:

$-y = x^3 - 2x + 1 \Rightarrow y = -x^3 + 2x - 1$

17. (B)

$1500 = 13.8n + 4.51 \Rightarrow n = \dfrac{1500 - 4.51}{13.8} = 108.369$

Therefore $\dfrac{108}{12} = 9$ dozens of items can be produced.

18. (E)

x + y must equal 90° therefore y can equal 1 or 3.

19. (E)

$g(2^3 - 1) - f((2-1)^3) = g(7) - f(1) = (7-1)^3 - (1^3 - 1) = 216$

20. (D)

$|AC| = \sqrt{64 + 16} = 4\sqrt{5}$

$f(\hat{A}) = \dfrac{1}{\sin A} = \dfrac{1}{\dfrac{4}{4\sqrt{5}}} = \dfrac{4\sqrt{5}}{4} = \sqrt{5}$

21. (C)

$y^2 + y = x^2 + x$

$y^2 + y - x^2 - x = 0$

$y^2 - x^2 + y - x = 0$

$(y - x)(y + x) + (y - x) = 0$

$(y - x)(y + x + 1) = 0$

which implies that y =x or y + x + 1 = 0. The equation represents two straight lines.

22 (C)

I. If a plane S is perpendicular to both planes, then one point is common to all three planes.
II. If the plane S also contains the line d but is different from the given planes P and Q then line d will be common to all three.

23. (E)

For any two similar solids ratio of areas is the square of the ratio of sides and ratio of volumes is the cube of the ratio of sides. Therefore ratio of sides equals $\dfrac{2}{\sqrt{5}}$ and ratio of volumes equals

$\left(\dfrac{2}{\sqrt{5}}\right)^2 = \dfrac{8}{5\sqrt{5}} = \dfrac{8\sqrt{5}}{25}$.

24. (A)

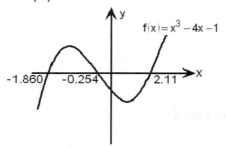

Greatest negative x intercept is -0.254.

25. (A)

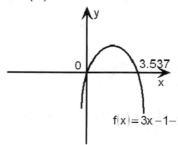

The graph of $f(x) = 3x + 1 - 2^x$ is given above and the values of x for which y is positive are $0 < x < 3.54$.

26. (A)

$f(0) = 0^2 + 0 \cdot m + n = 2 \Rightarrow n = 2$

$f(1) = 1^2 + m \cdot 1 + n = 0 \Rightarrow 1 + m + 2 = 0 \Rightarrow m = -3$

$f(x) = x^2 - 3x + 2 = 0 = (x-2)(x-1) \Rightarrow x_1 = 1$ and $x_2 = 2 = a$

27. (D)

$\sin(2x) = \sin x \Rightarrow 2\sin x \cdot \cos x = \sin x \Rightarrow \sin x = 0$ or $\cos x = \dfrac{1}{2}$

Therefore x can be 60°, 180°, or 360°.

28. (C)

4x-10=2 and 3x-4=E; therefore x=3 and $E = 3 \cdot 3 - 4 = 5$.

29. (D)

$\log_2 x^3 = 11 \Rightarrow x^3 = 2^{11} \Rightarrow x = \sqrt[3]{2^{11}} \cong 12.699$

30. (D)

Cosx is always positive whereas sinx can be positive or negative. Therefore we are in the first or the fourth quadrants.

31. (E)

N	2	3	4...	n
Number of points of intersection	1	3	6...	$\binom{n}{2}$

$\binom{n}{2} = \dfrac{n \cdot (n-1)}{2} = \dfrac{n^2 - n}{2}$

32. (B)
If we multiple both sides of the given equation by -1.5. we get, -3x+y=-4.5. Therefore we have two lines that coincide which gives infinitely many solutions.

33. (C)
$$x \cdot \frac{9.75}{100} - x \cdot \frac{8.50}{100} = \$18.75 \Rightarrow x \cdot 1.25 = \$1875$$
$$\Rightarrow x = \$1500$$

34. (C)
We omit the positive x portion of f(x); we reflect and duplicate the negative x portion of f(x).

35. (A)
$\sin^2 x > \cos x$ is not satisfied when x equals 6.18 radians.

36. (E)
$\sin 340° = \sin 200° = -\sin 20°$

37. (A)
|AC| − |BC| is a maximum when A, B and C are collinear.
Therefore slopes of AB and BC must be equal:
$$\frac{5-1}{2-8} = \frac{2-0}{8-p} \Rightarrow p = 12$$

38. (B)
For any three dimensional object we have V + F = E + 2 where V, F, and E are the numbers of vertices, faces and edges respectively. Therefore 5 + F = 8 + 2 and F = 5. The three dimensional object can be a square pyramid.

39. (D)
Center is (-15 + 7) / 2 = -4
Radius is (7 - -15) / 2 = 11.
$$|x - (-4)| < 11 \Rightarrow |x + 4| < 11$$

40. (B)
The new mean is 10 − 10 = 0 and the new standard deviation is 2 / 2 = 1

41. (C)
1st day 6 situations are possible
2nd day 5 situations are possible
3rd day 4 situations are possible
4th day 3 situations are possible
5th day 2 situations are possible. The probability
is: $\dfrac{6 \cdot 5 \cdot 4 \cdot 3 \cdot 2}{6 \cdot 6 \cdot 6 \cdot 6 \cdot 6} = \dfrac{5 \cdot 4 \cdot 3 \cdot 2}{6 \cdot 6 \cdot 6 \cdot 6}$

42. (C)
f(x) is first translated 1 unit to the right and we get f(x-1); then, it is reflected with respect to the y-axis and we get
f(-x-1); f(-x-1)=f(ax+b). So, a=-1 and b=-1

43. (D)
$$B(a,-b), C(-b,a), D(b,a) \Rightarrow |CD| = \sqrt{(-b-b)^2 + (a-a)^2} = |2b|$$

44. (C)
New number of people is n + 5m when the new price is a − mb. Therefore the new income is (a − mb)·(n + 5m).

45. (C)
$$\text{Amplitude} = \frac{y_{max} - y_{min}}{2} = \frac{4-1}{2} = 1.5$$

46. (B)
Calculate $\dfrac{\sin(2x^2)}{x^2}$ by replacing x with 0.00001; we get a value very close to 2.

47. (C)
$$\frac{(2\cos\theta)^2}{4} + \frac{y^2}{9} = 1 \Rightarrow \frac{4\cos^2\theta}{4} + \frac{y^2}{9} = 1$$
$y = \pm 3\sin\theta$ satisfies the above equation.

48. (D)
The other vertices are (0,1,0), (2,1,2), (0,-1,2) and (0,-1,0).

49. (C)
$$x^2 - 1 = 0 \Rightarrow (x-1)(x+1) = 0$$
$$x_1 = 1 \wedge x_2 = -1$$
The domain consists of all real numbers except 1 and -1.

50. (D)
The first few terms are 1, 2, 3, 5, 8, 13, 21, 34, 55, 89,... If O denotes an odd number and E denotes an even number then the sequence is O, E, O, O, E, O, O, E, O,... As it can be seen the O, E, O pattern continually repeats. So there are 10000/3·2 + 1 = 6667 odd numbers.

Model Test 11 – Solutions

1. (D)
$$\frac{2(3x+5y)}{4} = \frac{-m}{4} \Rightarrow m = -2(3x+5y) = -6x - 10y$$

2. (C)
$$\frac{1}{1 + \dfrac{1}{1+x}} = \frac{1}{\dfrac{1+x+1}{1+x}} = \frac{x+1}{x+2}$$

3. (E)
-3^{2x} is always negative so it can never be 1.

4. (B)
Midpoint of the line AB can either be in the first or the second quadrant or on the positive of the y − axis. It can not be on the x-axis.

5. (D)

In order to cut a wood into 10 pieces 9 cuts are needed. In order to cut the wood into 20 pieces 19 cuts are needed. So, if 90 seconds are required for 9 cuts, 190 seconds are required for 19 cuts.

6. (C)

Replace each x by x-1 in $f(x)=\begin{cases} 3x^2 & x \le 1 \\ 4 & x > 1 \end{cases}$.

$f(x-1)=\begin{cases} 3(x-1)^2 & x-1 \le 1 \\ 4 & x-1 > 1 \end{cases}$ which implies that

$f(x-1)=\begin{cases} 3(x-1)^2 & x \le 2 \\ 4 & x > 2 \end{cases}$.

7. (A)

$\dfrac{x}{\left(\dfrac{1}{x}\right)^2} < x \Rightarrow x^3 < x \Rightarrow x(x-1)(x+1) < 0.$

So, x<-1 or 0<x<1

8. (E)

There is no value that satisfies $\dfrac{9}{x+1}=0$. So x-1 is undefined.

9. (D) $x^2 + (y+2)^2 = 0 \Rightarrow x = 0$ and $y+2 = 0 \Rightarrow y = -2$
Therefore y = x-2.

10. (E)

$\left.\begin{array}{l} 2x - 2y = 4 \Rightarrow x - y = 2 \\ 3x + 3y = -9 \Rightarrow x + y = -3 \end{array}\right\}$

$x^2 - y^2 = (x-y)\cdot(x+y) = 2\cdot(-3) = -6$

11. (A)

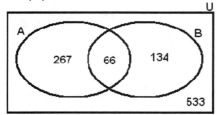

A checks every 3rd test, totally 333 tests. B checks every 5th test, totally 200 tests. A and B both check every 15th test, a totally 66 tests. Therefore the number of tests that not checked by any of the examiners is: 1000-(267+66+134)=533

12. (D)

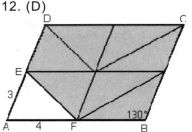

$m(A\hat{B}C) = 130°, m(D\hat{A}B) = 180° - 130° = 50°$

$A(BCDEF) = 7 \cdot A(E\hat{A}F) = 7 \cdot \frac{1}{2} \cdot 3 \cdot 4 \cdot \sin 50° = 32.17 \approx 32$

13. (C)

The profit from a single item is $\$\left(\dfrac{a}{b} - \dfrac{b}{a}\right)$. So,

the rate of the profit is $\dfrac{\left(\dfrac{a}{b} - \dfrac{b}{a}\right) \cdot 100}{\dfrac{b}{a}}\%$

14. (B)

I and V are functions from A to B.

15. (A)

If $a \cdot c < 0$ then a and c have opposite signs. Therefore if a is positive then c is negative or if a is negative then c is positive. This implies that if the curve opens upward, it must have a negative y intercept as in I or if the curve opens downward, it must have a positive y intercept.

16. (D)

$\dbinom{10}{1} \cdot \dbinom{20}{2} = 10 \cdot \dfrac{20 \cdot 19}{2} = 1900$

17. (E)

$2,4,6,8 \Rightarrow 4 \cdot 1 = 4$

$10,12,14,..,98 \Rightarrow \left(\dfrac{98-10}{2} + 1\right) \cdot 2 = 90$

$100,102,..,1000 \Rightarrow \left(\dfrac{1000-100}{2} + 1\right) \cdot 3 = 1353$

4+90+1353=1447

18. (C)

Let e be the identity element for the operation @. $x@e = x + e - 2xe = x$ must be satisfied therefore $e(1-2x) = 0 \Rightarrow e = 0$.

19. (D)

The equation has distinct roots, therefore discriminant is positive. The roots are integral, therefore discriminant is a perfect square.

20. (D)
Radius is 10 feet = 120 inches.

Area of the garden is $\dfrac{\pi \cdot 120^2}{2}$. Number of roses

needed is $\dfrac{\pi \cdot 120^2}{2 \cdot 15}$. Number of packs needed is

$\dfrac{\pi \cdot 120^2}{2 \cdot 15 \cdot 8}$. Total cost is $\dfrac{\pi \cdot 120^2}{2 \cdot 15 \cdot 8} \cdot \$7 = \$1{,}319$.

21. (C)
1, 2, 3,..., 9: totally 9 digits
The rest are two digit numbers.

81-9=72 digits left.
$72/2 = 36^{th}$ two digit number's units digit is the digit we are looking for.
10, 11, ..., 45: There are totally 36 two digit numbers here. 5 is the answer.

22. (A)

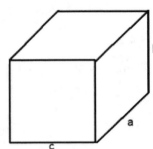

$ab = 3$
$bc = 5$
$ac = 15$
Multiply the equations side by side and we get

$a^2 b^2 c^2 = 225 \Rightarrow abc = 15 =$ Volume

23. (B)
5 people started with Civil and ended with Electrical.

24. (C)
105 – 55 = 50 people dropped out of Computer Engineering and this was the greatest drop out.

25. (B)
66+68+105+41+93=373 in the beginning.
68+77+76+47+113=381 in the end.
So enrollment increased by 381−373=8.

26. (B)
$x^4 + x^2 + 1 = 0$

$x^4 + x^2 + 1 = 0 \Rightarrow x^4 + 1 = -x^2 \Rightarrow \dfrac{x^4 + 1}{x^2} = -1$

$\Rightarrow x^2 + \dfrac{1}{x^2} = -1 \left(x^2 + \dfrac{1}{x^2} \right)^2 = (-1)^2 \Rightarrow x^4 + 2 + \dfrac{1}{x^4} = 1$

$\Rightarrow x^4 + \dfrac{1}{x^4} = -1$

27. (E)

$r = 6 - 3\sqrt{\dfrac{g_6}{g_3}} = 3\sqrt{\dfrac{4}{32}} = \dfrac{1}{2}$

$g_3 = g_1 \cdot \left(\dfrac{1}{2} \right)^{3-1} \Rightarrow 32 = g_1 \cdot \left(\dfrac{1}{2} \right)^2 \Rightarrow g_1 = 128$

$g_{11} = 128 \cdot \left(\dfrac{1}{2} \right)^{11-1} = 2^7 \cdot 2^{-10} = 2^{-3}$

$g_{15} = 128 \cdot \left(\dfrac{1}{2} \right)^{15-1} = 2^7 \cdot 2^{-14} = 2^{-7}$

$g^{11} + g^{15} = \dfrac{1}{8} + \dfrac{1}{128} = \dfrac{17}{128}$

28. (C)
Three cases are possible as indicated in the figure below:

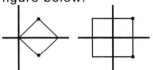

29. (A)
Figure corresponds to the equation given in A.

30. (B)
$2\sin\theta \cdot \cos\theta - \sin^2\theta - \cos^2\theta - 2\sin\theta \cdot \cos\theta$
$= -\left(\sin^2\theta + \cos^2\theta \right) = -1$

31. (E)
The points (1, 2), (1, -2), (-2, 1) and (-1, -2) satisfy the given relation. Therefore answer is E.

32. (A)
The number of complex roots of a real polynomial must be even. Therefore a 7th degree polynomial cannot have 0 real (and 7 complex) roots.

33. (C)

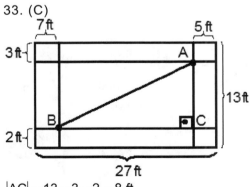

$|AC| = 13 - 3 - 2 = 8$ ft

$|BC| = 27 - 7 - 5 = 15$ ft

$AB = \sqrt{8^2 + 15^2} = 17$ ft

(233)

34. (C)

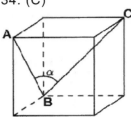

Since |AC| = |AB| = |BC|, triangle ABC is equilateral.

So, $m(A\hat{B}C) = \alpha = 60°$

35. (C)

720 square inches = 5 square feet. Area of the region that consists of the points (x,-2y) is twice that of the original one and it equals 10 square feet.

36. (E)

$\dfrac{2}{3} = \dfrac{-m}{2} \neq \dfrac{3}{n}$ Therefore m=-4/3 and n≠9/2.

37. (B)

$x = \dfrac{my+n}{py-q} \Rightarrow xpy - xq = my + n \Rightarrow xpy - my = xq + n$

$\Rightarrow y(xp-m) = xq+n \Rightarrow y = f^{-1}(x) = \dfrac{qx+n}{px-m}$

38. (C)

It's a hyperbola and it has the symmetries given in I, II and III.

39. (D)

$\left.\begin{array}{l} y = -x^2 + 2x \\ y = mx + 9 \end{array}\right\} -x^2 + 2x = mx + 9$

$\Rightarrow x^2 + (m-2)x + 9 = 0$

$\Delta = (m-2)^2 - 4 \cdot 1 \cdot 9 = 0 \Rightarrow (m-2)^2 = 36$

(i) m-2=6 ⟹ m=8

(ii) m-2=-6 ⟹ m=-4

40. (C)

1. $(x-3)^2$; 2. $(x-3)^2 - 2$;

3. $(-x-3)^2 - 2 = (x+3)^2 - 2$

41. (D)

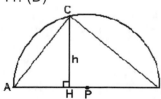

Please note that ACB is a right triangle whose right angle is at C.

$CA = 2R\cos\alpha \Rightarrow h = CH = CA\sin\alpha = 2R\cos\alpha\sin\alpha$

42. (D)

r is the radius of one of the small spheres and R is the radius of the big sphere.

$8 \cdot \dfrac{4}{3}\pi r^3 = \dfrac{4}{3} \cdot \pi \cdot R^3 \Rightarrow \dfrac{R^3}{r^3} = 8 \Rightarrow \dfrac{R}{r} = 2$.

The ratio of the area of the big sphere to the total area of the small spheres is

$\dfrac{4\pi R^2}{8 \cdot 4\pi r^2} \Rightarrow \dfrac{R^2}{8r^2} = \dfrac{1}{8} \cdot \dfrac{R^2}{r^2} = \dfrac{1}{8} \cdot 4 = \dfrac{1}{2}$

43. (D)

$\log_e(y^2) = x \Rightarrow y^2 = e^x \Rightarrow \sqrt{y^2} = \sqrt{e^x} \Rightarrow |y| = e^{x/2}$

$\Rightarrow y = \mp ex/2$

44. (B)

$\dfrac{6}{7} < \dfrac{a}{b} < \dfrac{7}{8} \Rightarrow \dfrac{12}{14} < \dfrac{a}{b} < \dfrac{14}{16}$

Therefore minimum integer value of b is 15.

45. (B)

$(i-1)^2(2i+1) = (-1-2i+1)(2i+1) = -4i^2 - 2i = 4 - 2i$

46. (E)

The contrapositive of the statement p⟹q is q'⟹p'. "If Cem takes the test then he will score 800" is equivalent to the statement "If he did not score 800, then he did not take the test".

47. (A)

$195 = 25Q + 5N \Rightarrow N = 39 - 5Q$

The possibilities are

Q = 1 and N = 34   Q = 2 and N = 29
Q = 3 and N = 24   Q = 4 and N = 19
Q = 5 and N = 14   Q = 6 and N = 9
Q = 7 and N = 4

Therefore he has at least 11 coins.

48. (E)

$(x-2y)^{-4}$ has infinitely many terms.

49. (D)

$\mathbf{B}_{4\times2} \times \mathbf{A}_{2\times4} = (\mathbf{BA})_{4\times4}$

50. (C)

The polynomial given by I may also have complex zeros. Therefore it may have more than 3 zeros.

Model Test 12 – Solutions

1. (B)

$\dfrac{a}{c} = \dfrac{b/3}{b/4} = \dfrac{4}{3}$

2. (C)

For the same positive integer value of x, $\dfrac{2}{3x+1}$ is the least.

3. (E)

If we multiply both sides of the equation by 100, we get 50x+10x-15x=180 ⟹45x=1800⟹x=40

4. (B)
$h(a, b, c) = (2a, 3b, -2c)$
$g(h(a, b, c,)) = \left((2a)^2, (3b)^3, -2c\right) = \left(4a^2, 27b^3, -2c\right)$

5. (D)

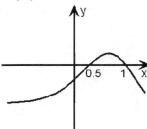

$f(2x) = 2^{2x} - 3^{2x-1} = 1$

If we plot the graph of the function

$y = 2^{2x} - 3^{2x-1} - 1$ with a graphing calculator we will observe that one of the zeros of the function is 1.

6. (E)
$3 \cdot f(3) = 3\left(3\sqrt{3} + \sqrt{3.3}\right) = 9\sqrt{3} + 9 = 9\left(\sqrt{3} + 1\right)$

7. (C)
Center of the circle: $\left(\dfrac{-6+7}{2}, \dfrac{8+24}{2}\right) = \left(\dfrac{1}{2}, 16\right)$

8. (A)
$x^2 + 4xi - 4 = y - 6i$
$4x = -6 \Rightarrow x = -1.5$ and
$x^2 - 4 = y = -1.5^2 - 4 = -1.75$

9. (C)

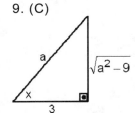

$\sin x = \begin{cases} +\dfrac{\sqrt{a^2 - 9}}{9}, & \text{If } 0 < x < \frac{\pi}{2} \\ -\dfrac{\sqrt{a^2 - 9}}{9}, & \text{If } \frac{\pi}{2} < x < 0 \end{cases}$

10. (A)
$(3 + x)(51 + x) = (15 + x)^2$
$153 + 3x + 51x + x^2 = 225 + 30x + x^2$
$\Rightarrow 24x = 72 \Rightarrow x = 3$

11. (B)
$g(h(2)) = g(3) = -1$

12. (B)
If $y = |\sin(2x)|$ is graphed in the radian mode, it will be observed that its maximum and minimum values are 1 and 0 respectively. Therefore $0 \le y \le 1$.

13. (C)
$P_{2000} = 133,000 \cdot (1 - 0.03)^{2010 - 2000} =$
$133,000 \cdot (0.97)^{10} = 98,077.40 \approx 98,100$

14. (D)
center of the circle:(1,1)

$r = \sqrt{(-3-1)^2 + (2-1)^2} = \sqrt{16+1} = \sqrt{17}$

For only (5, 0) the distance between the selected point and the center of the circle is $\sqrt{17}$.

15. (E)

p	2	4	6	8
3p+1	7	13	19	25

We do not need to calculate 3p+1 for p = 0. 7, 13 and 19 are prime but 25 is not.

16. (D)

If $\cos\theta = 0.35 = \dfrac{35}{100} = \dfrac{7}{20}$

$z = \sqrt{20^2 - 7^2}$ $\dfrac{z}{y} = \dfrac{\sqrt{20^2 - 7^2}}{7} \cong 2.676$

17. (D)
$x^4 + x^2 + 1 > 0$; therefore, $x + 2 = 0$, and $x = -2$
For $x = -2$
$(-2 - 2)\left((-2)^3 + (-2)^2 + 1\right) = -4 \cdot (-8 + 4 + 1) = 12$

18. (E)
$f(x) = 2 \cdot 2x - 1 \Rightarrow 2x = \dfrac{f(x) + 1}{2} \Rightarrow g(x) = \dfrac{f(x) + 1}{2} + 1 = \dfrac{f(x) + 3}{2}$

19. (A)
$\dfrac{4 + 2 - 1}{8x} = \dfrac{1}{4} \Rightarrow 20 = 8x \Rightarrow x^{-1} = \dfrac{8}{20} = 0.4$

20. (D)
$-2 < 3x - 2 \le 4 \Rightarrow 0 < 3x \le 6 \Rightarrow 0 < x \le 2$

21. (B)
$t = 19.6 \cdot (4500)^{\frac{1}{3}} \cong 323.58 \,\text{hours} = 13.48 \,\text{days}$

22. (C)

5R 4G → 6R 3G   Probability = $\dfrac{6}{9} \cdot \dfrac{5}{8} \cdot \dfrac{4}{7} = \dfrac{5}{21}$

23. (E)
A horizontal line has a equation of the from y=a where a is any real number. So, there must be any variable other than y.

24. (E)
Correct answer is given in E.

25. (B)
$$\frac{\binom{7}{2}\cdot\binom{6}{1}}{\binom{5+6+7}{3}}=\frac{\frac{7\cdot6}{2}\cdot6}{\binom{18}{3}}=\frac{7\cdot3\cdot6}{\frac{18\cdot17\cdot16}{3\cdot2}}=\frac{21}{136}$$

26. (C)
$$\left.\begin{array}{l}f(-1)=-2+\dfrac{B}{2}\\[2mm]f(1)=2-\dfrac{B}{2}\end{array}\right\}\quad\text{Therefore }\dfrac{B}{2}-2=8-2B\Rightarrow B=4$$
and $\dfrac{B}{2}-2=4\cdot\left(2-\dfrac{B}{2}\right)$

27. (C)
$y=x^2+3$. So, for $x=6$ we get $y=6^2+3=39$

28. (A)
$\tan^2 x-\sec^2 x=\tan^2 x-\left(1+\tan^{2x}\right)=-1$

29. (B)
The third angle will be 180-(110+40)=30
By the sine rule
$$\frac{30}{\sin 40}=\frac{x}{\sin 30}\Rightarrow x=\frac{30\cdot\sin 30}{\sin 40}\cong 23.335$$

30. (D)
$$\lim_{x\to\infty}\frac{-3x+1}{-x+1}=\frac{-3}{-1}=3$$

31. (A)

If we reflect the given graph of the function f(x) with respect to the line y=x, we'll get the graph on the left hand side.

32. (C)
The numbers 10000, 10000 and 10000 are closest to each other than the others. So, they have the least standard deviation.

33. (B)
$$\frac{2\cdot10+4\cdot20+6\cdot30+6\cdot40+8\cdot50+10\cdot60+8\cdot70+4\cdot80+4\cdot90+2\cdot100}{2+4+6+6+8+10+8+4+4+2}$$
=54.8

34. (A)
$$\left.\begin{array}{l}f(x)=ax+b\\f(x+1)=ax+a+b\end{array}\right\}f(x+1)-f(x)=a=6$$
$$\left.\begin{array}{l}f^{-1}(x)=\dfrac{x-b}{a}\\[2mm]f^{-1}(x+1)=\dfrac{x+1-b}{a}\end{array}\right\}f^{-1}(x+1)-f^{-1}(x)=\dfrac{1}{a}=\dfrac{1}{6}$$

35. (E)
$\log_5(5-x)=0\Rightarrow5-x=5^0\Rightarrow5-1=x=4$

36. (C)
$x=e^{2y}\Rightarrow\ln x=2y\Rightarrow f^{-1}(x)=\dfrac{1}{2}\cdot\ln x$
$f^{-1}(4)=\dfrac{1}{2}\cdot\ln 4\cong 0.6931$

37. (E)
In Case 1 x = y. In Case 2 x >y.
In Case 3 x < y. Therefore I and III are correct.

38. (E)
f has a numerator with a factor of x-3.
$x+1=0\Rightarrow x=-1$ is the vertical asymptote. $y=\dfrac{2}{1}=2$
is the horizontal asymptote.
$$f(x)=\frac{2x-6}{x+1}=\frac{2(x-3)}{x+1}$$

39. (D)
A: 2x + 3y + 4z = 5
B: 2x + 3y + 4z = 10/3
C: 2x + 3y + 4z= -5/2
A, B and C all represent parallel planes and B is between A and C.

40. (B)
If we plot the graph of the function
$y=2x^2-4x-5$ with a graphing calculator, we will observe that the zeros are at -0.87 and 2.87

41. (D)
A: $\dfrac{\cos x}{3}$ has a period of T.

B: $\cos\left(\dfrac{x}{3}\right)$ has a period of 3T

C: $\cos\left(x+\dfrac{1}{3}\right)$ has a period of T

D: $\cos(-3x)$ has a period of T/3

E: $3\cos x$ has a period of T

42. (A)
$$\frac{b}{a}=\frac{b^1}{b^x}=b^{1-x}$$

43. (C)
6 seniors out of 12 can be selected in $\binom{12}{6}$
ways. 4 juniors out of 8 can be selected in $\binom{8}{4}$
ways. So, the result is $\binom{12}{6}\cdot\binom{8}{4}$

(236)

44. (C)
When we add II and III side by side, we get
$7\sin^2 x + 7\cos^2 x = 7$ which is equivalent to
$\sin^2 x + \cos^2 x = 1$. Therefore II and III have the
same solution set.

45. (D)
$|5 \cdot e^{(-0.3x)} \cdot \sin(x) + 2.4| = 2.3$
$y = |5 \cdot e^{(-0.3x)} \cdot \sin(x) + 2.4| - 2.3 = 0$
When we plot the graph of
$y = |5 \cdot e^{(-0.3x)} \cdot \sin(x) + 2.4| - 2.3 = 0$
we will observe that the greatest x intercept is
at x = 11.80.

46. (A)
$h(t) = 20t - 5t^2$

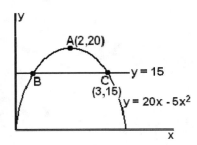

The information given in the graph can be obtained using a graphing calculator. Please note that y denotes height and x denotes time. Answer is 3-2=1.

47. (C)
$\frac{6}{16} = \frac{r}{4} \Rightarrow r = 1.5$ So, $V = \frac{1}{3} \cdot \pi \cdot 1.5^2 \cdot 6 = 14.137$

48. (C)
I. 1
II. $1 + \tan^2 x - \tan^2 x = 1$
III. $\frac{1}{\cos x} \cdot \frac{1}{\sin x} = \frac{1}{\cos x \cdot \sin x} \neq 1$

49. (E)

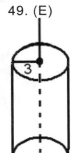

50. (B)
$l_1 = k \cdot \frac{\cos\theta}{d^2}$
d is doubled whereas $\cos\theta$ does not change.
$l_2 = k \cdot \frac{\cos\theta}{(2d)^2} = k \cdot \frac{\cos\theta}{4d^2}$. Therefore $\frac{l_1}{l_2} = \frac{4}{1} \Rightarrow l_2 = \frac{l_1}{4}$

Model Test 13 – Solutions

1. (E)

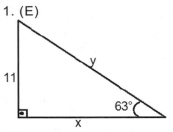

$\frac{11}{x} = \tan 63° \Rightarrow x = 11 / \tan 63° \cong 5.6$
$\frac{11}{y} = \sin 63° \Rightarrow y = \frac{11}{\sin 63°} = 12.35$
Perimeter $= 11 + 5.6 + 12.35 \cong 56.8$

2. (A)
$x + 4y - 6z = 5$
$2x - 2y + 3z = -10$
Multiply the second equation by 2.
$x + 4y - 6z = 5$
$4x - 4y + 6z = -20$
Add the equations above side by side.
$5x = -15$
$x = -3$

3. (D)
$a \cdot b = 2c - 1 \Rightarrow a \cdot b = \text{odd}$
$\left. \begin{array}{l} a \to \text{odd} \\ b \to \text{odd} \end{array} \right\} \text{odd} + \text{odd} = \text{even}$

4. (C)
$ax = by$ and $cy = dz$
If y is eliminated then $\frac{ax}{b} = \frac{dz}{c}$. Therefore
$\frac{x}{z} = \frac{bd}{ac}$ and $\frac{ex}{z} = \frac{bde}{ac}$

5. (E)
$\sin 20° = -\sin 340° \Rightarrow x = 340°$

6. (A)
$f(a+b) = f(a) + (b)$; $f(0+0) = f(0) + f(0)$
$f(0) = 2f(0)$ therefore $f(0) = 0$

7 (A)
If the graph of the function $y = \pi^x - \pi - x$ is
plotted, its zero will be observed as -3.11.

8 (E)
In any case $\frac{m-n}{mn}$ is a rational number and
every rational number is a repeating decimal.
Therefore E is false.

9. (C)
If y = 0 then x = -8; therefore all such lines pass
through (-8, 0).

10. (D)

$x = -3 \Rightarrow y = -3 \cdot -3 + a = 9 + a$

$x = -7 \Rightarrow y = -3 \cdot -7 + a = 21 + a$

Shaded region is a trapezoid whose area is

given by $\left(\dfrac{21 + a + 9 + a}{2}\right) \cdot 4 = 84$

$30 + 2a = 42 \Rightarrow a = 6$

11. (C)

$\sec^2 x - 3\tan x + 1 = 0$

$\sec^2 x - 1 - 3\tan x + 1 = -1$

$\tan^2 x - 3\tan x + 1 = -1$

$\tan^2 x - 3\tan x + 2 = 0$

$(\tan x - 1)(\tan x - 2) = 0$

Therefore $\tan x = 1$ or $\tan x = 2$.

12. (C)

Center C and vertex A have the same y coordinate that equals 2. If $y = 2$ then $y = 0.75x - 4$ will imply that $x = 8$. So the center is at $(8, 2)$ and $AC = 4 =$ semi transversal axis length. On the other hand if $x = 4$ then $y = -1$; therefore $(4, -1)$ is the lower left hand corner of the rectangle. Therefore semi conjugate axis length is 3 and the equation of the hyperbola is

$\dfrac{(x - 8)^2}{16} - \dfrac{(y - 2)^2}{9} = 1$.

13. (A)

If $y = -\dfrac{1}{2}(\sin x \cdot \cos x) + 1$ then it can be observed

that $y_{\max} = 1.25$ and $y_{\min} = 0.75$.

Amplitude $= \dfrac{y_{\max} - y_{\min}}{2} = \dfrac{1.25 - 0.75}{2} = 0.25 = \dfrac{1}{4}$.

14. (B)

$\cos(\pi - \theta) = -\cos\theta = -0.82$

15. (D)

1,2,3,4,5,6,7,8,9,10,11,12

Median $= \dfrac{6 + 7}{2} = 6.5$ and it is not an

integer.

Mean = 6.5; therefore median equals mean. Moreover, there is no mode in this set.

16. (C)

$a_1 = 5$

$a_2 = 7$

$r \rightarrow$ common ratio

$a_2 = a_1 \cdot r \Rightarrow r = \dfrac{7}{5}$

$a_{30} = a_1 \cdot r^{29}$

$a_{30} = 5 \cdot \left(\dfrac{7}{5}\right)^{29} \approx 86{,}434$

17. (E)

$f(0,8) = (3 \cdot 0.8)^{(2 \cdot 0.8)^{0.8}} = (2.4)^{(1.6)^{(0.8)}} = (2.4)^{1.46} \approx 3.58$

18. (B)

for $x = \pi$

$-1 = \cos\pi + i\sin\pi = e^{i\pi}$

$(-1)^i = \left(e^{i\pi}\right)^i = e^{i^2\pi} = e^{-\pi}$

19. (A)

If the smallest number is removed median will either stay the same or increase.

20. (C)

$\dfrac{y + 1}{4} = \cos(3\theta + \pi)$

$-1 \leq \cos(3\theta + \pi) \leq 1$

$-1 \leq \dfrac{y + 1}{4} \leq 1$

$-4 \leq y + 1 \leq 4$

$-5 \leq y \leq 3$

21. (C)

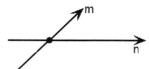

The point of intersection of m and n stays fixed.

22. (E)

There are two spheres that intersect to give a circle on which there are infinitely many points.

23. (A)

$\dfrac{\dfrac{x^2 - 1}{x - 1}}{x} = \dfrac{x \cdot (x^2 - 1)}{(x - 1)} = \dfrac{x \cdot (x - 1) \cdot (x + 1)}{(x - 1)} = x^2 + x$

The above expression will be valid only if $x \neq 1$ and $x \neq 0$.

24. (D)

$x = 1$ is a vertical asymptote so domain $\Re - \{1\}$ and the range is $y < -1$ or $y > 0$.

25. (C)

Common ratio is $-(x + 2) = -x - 2$. If $-1 < r < 1$

then sum is given by $\dfrac{a_1}{1 - r} = \dfrac{1}{1 - (-x - 2)} = \dfrac{1}{x + 3}$.

So $-1 < -x - 2 < 1$ and therefore $-3 < x < -1$.

26. (B)

$\dfrac{26 \quad 25 \quad 24 \quad 23 \quad 22 \quad 21}{} = P(26,6)$

27. (E)

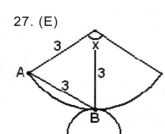

Perimeter of the base is $2\pi = \dfrac{2\pi \cdot 3 \cdot x}{360°}$.

Therefore $x = 120°$ and $\dfrac{x}{2} = 60°$; which leads to AB = 3.

28. (B)

A·X = B

$$\begin{bmatrix} 2 & -1 & 0 \\ 1 & 2 & -3 \\ -1 & 1 & 1 \end{bmatrix} \cdot \begin{bmatrix} x \\ y \\ z \end{bmatrix} = \begin{bmatrix} 4 \\ 3 \\ -3 \end{bmatrix}$$

Therefore B has 3 rows and 1 column.

29. (A)

5x3=15 cuts are needed.

30. (A)

$\left(e^a\right)^n = e^{an}$ therefore f(x) = e^x.

31. (A)

The coefficient a must be non zero otherwise equation will not be quadratic and it will not be possible for it to have two real and distinct solutions.

32. (C)

$\tan(x) = \tan(x + \pi)$ therefore f(x) = tan(x).

33. (C)

If we replace x by a value that is very close to 1, such as 0.99999 and 1.00001, we will see that $\dfrac{x-1}{\ln x}$ is very close to 1.

34. (B)

If f(x) opens downward and does not intersect the x axis then f(x) and |f(x)| intersect at no points.
If f(x) opens downward intersects the x axis at a single point then f(x) and |f(x)| intersect at 1 point.

If f(x) opens upward and does not intersect the x axis then f(x) and |f(x)| intersect at infinitely many points.

35. (A)

$4 = \log_a 100 \Rightarrow a^4 = 100 \Rightarrow a = 3.16$

$b = \log_a 0.1 \Rightarrow a^b = 0.1$

$3.16^b = 0,1 \Rightarrow b = \dfrac{\log 0.1}{\log 3.16} \approx -2$

36. (D)

BB+RR+GG

$\dfrac{7}{24} \cdot \dfrac{6}{23} + \dfrac{8}{24} \cdot \dfrac{7}{23} + \dfrac{9}{24} \cdot \dfrac{8}{23} = \dfrac{9 \cdot 8 + 8 \cdot 7 + 7 \cdot 6}{24 \cdot 23}$

37. (E)

$r^2 = (-6)^2 + 6^2 \Rightarrow r = 6\sqrt{2}$

$\theta = \dfrac{-5\pi}{4}$ since the point is in the third quadrant and it makes $\dfrac{\pi}{4}$ radians with the horizontal.

38. (A)

$x^2 - 3 = x \Rightarrow x^2 = x + 3$

We can replace x^2 by x+3 every time it appears.

$x + \dfrac{9}{x^2} + 3 = x + 3 + \dfrac{9}{x+3} = \dfrac{(x+3)^2 + 9}{x+3} = \dfrac{x^2 + 6x + 9 + 9}{x+3}$

$= \dfrac{x + 3 + 6x + 9 + 9}{x+3} = \dfrac{7x + 21}{x+3} = 7$

39. (C)

f(x) shifted 1 unit left to get f(x+1), then 1 unit up to get f(x+1)+1.

40. (C)

f(x) and –f(x) intersecting at a single point implies that f(x) is tangent to the x axis, therefore its discriminant is zero which implies C.

41. (D)

An even function is not symmetric with respect to the x axis.

42. (C)

Distance between A and D
AD = 130 and AE = 170 therefore
DE = 40 = y.
AE = 170 and BE = 90 therefore
AB = 80 = x.
x + y = 80 + 40 = 120

43. (C)

New price is

$\$300 \cdot 0.8 \cdot 1.12^3 + \$300 \cdot 0.2 = \$397$

44. (B)

First calculate tan250° in degree mode to get 2.75. Then calculate sin(2.75) in radian mode and the answer is -0.923.

45. (E)

x cannot be 0, 1 or -1.

46. (D)

$\dfrac{2\pi}{\dfrac{\pi}{4}} = 8$ and the figure is an octagon.

47. (E)
The function can be graphed and the zeros will be observed to be -2, -1 and 2 and only one of them is odd.

48. (D)
600 miles for 4 tires makes a total of 2400 miles. If she would like to share it equally among each tire she will drive with each tire for $\frac{2400}{5} = 480$ miles.

49. (A)
In order to obtain f(−x) reflect f(x) with respect to the y−axis. Then reflect f(−x) with respect to the x−axis in order to obtain −f(−x).

50. (D)
The set given in II does not have a greatest element.

Model Test 14 – Solutions

1. (D)
$x = t^5 \Rightarrow t = \sqrt[5]{x} \Rightarrow y = 2 \cdot \left(\sqrt[5]{x}\right)^3 + 1 = 2 \cdot \sqrt[5]{x^3} + 1$

2. (E)
$\dfrac{x}{y} + \dfrac{y}{x} - 1 = \dfrac{x^2 + y^2 - xy}{xy}$

3. (E)
$\dfrac{3}{4}x^2 = 4 \Rightarrow x^2 = \dfrac{16}{3} \Rightarrow x = \mp 2 \cdot 309 \Rightarrow x^3 = \mp 12.32$

4. (E)

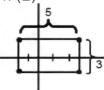

Perimeter=2(5+3)=16

5. (A)
$\dfrac{\tan\theta}{\sec\theta} = \dfrac{\frac{\sin\theta}{\cos\theta}}{\frac{1}{\cos\theta}} = \dfrac{\sin\theta}{\cos\theta} \cdot \cos\theta = \sin\theta = 0.65$

6. (E)
$\sec\theta = \dfrac{1}{\cos\theta} = 1.67 \Rightarrow \cos\theta = \dfrac{1}{1.67} = \dfrac{23.45}{x}$
$\Rightarrow x = 23.45 \cdot 1.67 = 39.16$

7. (D)
$g(f(5)) + f(f(8)) = g(5^2 + 2) + f(\sqrt[3]{8} + 1)$
$= g(27) + f(3) = \sqrt[3]{27} + 1 + 3^2 + 2 = 4 + 9 + 2 = 15$

8. (E)
$(x - 3)^2 + (y + 2)^2 = 25$
$(6 - 3)^2 + (y + 2)^2 = 25$
$9 + (y + 2)^2 = 25 \Rightarrow y + 2 = \pm 4$
$y = 2 \text{ or } y = -6$
However y=2 because point is in the first quadrant.

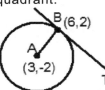

The slope of line AB is $\dfrac{2 - -2}{6 - 3} = \dfrac{4}{3}$ and the slope of line T is $-\dfrac{3}{4}$.
Line T passes through B(6, 2) therefore its equation is given by
$y - 2 = -\dfrac{3}{4}(x - 6)$.

9. (B)
Slope of line=-3; therefore slope of line m is $\dfrac{-1}{-3} = \dfrac{1}{3}$. Answer is B.

10. (D)
Range is the set of all y values and it is given by $-2 \leq y \leq 2$

11. (C)
$\dfrac{1 + \cot^2\theta}{\cot^2\theta} = \dfrac{1}{\cot^2\theta} + 1 = \tan^2\theta + 1 = \sec^2\theta$

12. (E)
Figure represents left hand portion of the unit circle therefore
$x^2 + y^2 = 1 \Rightarrow x^2 = 1 - y^2 \Rightarrow x = -\sqrt{1 - y^2}$

13. (C)
Answer is C. The function given in C has two zeros at x=2 and x=-3; a hole at x=1 and a vertical asymptote at x=-5

14. (E)
f(x) − k has a single zero for all k; therefore for k = 0 it intersects the x axis exactly once.

15. (C)
$\dfrac{f(8) \cdot (f(5) - f(4))}{f(4)} = \dfrac{6 \cdot (3 - 2)}{2} = \dfrac{6}{2} = 3 = f(5)$

16. (E)
$\dfrac{4}{3}\pi R^3 = \pi R^2 h \Rightarrow \dfrac{4R}{3} = h \Rightarrow h = \dfrac{4 \cdot 5}{3} = \dfrac{20}{3}$
Total area of the cylinder is
$= 2\pi R^2 + 2\pi Rh = 2\pi R(R + h) = 2 \cdot \pi \cdot 5\left(5 + \dfrac{20}{3}\right) = 366.52$

17. (D)
By the sine rule
$$\frac{8\cdot1}{\sin85°}=\frac{AC}{\sin55°}\Rightarrow AC=6.66$$

18. (A)
New median cannot be less than the original median since the smallest number is removed from the set.

19. (D)

$$x^2-2x-3<0$$
$$(x-3)(x+1)<0\Rightarrow-1<x<3$$
$$\Rightarrow-2<f(x)\le2$$

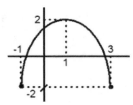

20. (C)
If $y=\left|x^2-4\right|$ is plotted for $-3\le x\le1$ then it will
be observed that the range is $0\le y\le5$.

21. (D)

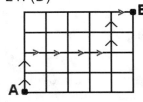

U: Up
R: Right
Indicated path is UURRRRUUR. The number of all such paths are the number of all permutations of the word UURRRRUUR

$$\Rightarrow\frac{9!}{5!\cdot4!}=C(9,4).$$

22. (B)
$$a_3=ar^2=4$$
$$a_6=ar^5=-108$$
$$\frac{ar^5}{ar^2}=\frac{-108}{4}=-27$$
$$r^3=-27\Rightarrow r=-3$$
$$a_2=\frac{a_3}{r}=\frac{4}{-3}=-\frac{4}{3}$$

23. (A)
Plot $y=\frac{\cos x}{\sin x}$ and $y=\sin x$ in the given intervals.
You will see that A is the correct choice.

24. (B)
$f(x)=(3x)^2$ is not a one to one function, therefore it does not pass the horizontal line test and it does not have an inverse.

25. (B)
$$x^3-4x^2+4x\neq0$$
$$x\cdot(x^2-4x+4)\neq0$$
$$x\cdot(x-2)^2\neq0$$
$$x\neq0\wedge x\neq2$$

26. (E)
$$e^{\sin(-\theta)}\cdot e^{-\sin(\theta)}=e^{-\sin\theta-\sin\theta}=e^{-2\sin\theta}$$

27. (E)
I is false because $\left|f(x)\right|=0$ when x=0. The rest is correct.

28. (B)
$$(x^2+1)\cdot(2x-1)+(x-x^2)\cdot2x=0$$
$$2x^3-x^2+2x-1+2x^2-2x^3=0$$
$$x^2+2x-1=0$$
By the quadratic formula $x=-1\mp\sqrt2$

29. (A)
$$f(g)=\frac{2(x+2)}{x-r}$$
$$-r=2\Rightarrow r=-2$$

30. (A)

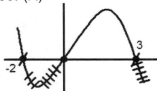

solution set is
$-2\le x\le0$ or $x\ge3$

31. (A)
$$(x-0)^2+(y-0)^2=1^2$$
$$x^2+y^2=1\Rightarrow x^2+y^2-1=0$$

32. (D)
The graph of the function is given below.

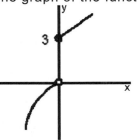

Range is given correctly in D.

33. (D)
$$0.5x^2=3\Rightarrow x^2=6$$
$$x>0\Rightarrow x=\sqrt6=2.45$$

34. (A)
$$x = \frac{1}{y-1}$$
$$xy - x = 1$$
$$xy = x + 1$$
$$y = \frac{x+1}{x} = f^{-1}(x) \Rightarrow f^{-1}(x) \text{ is undefined when } x=0$$

35. (C)
$$a_7 + 3d = a_{10}$$
$$4 + 3d = 13$$
$$3d = 9 \Rightarrow d = 3$$
$$a_{72} = a_7 + 65 \cdot 3 = 4 + 195 = 199$$

36. (C)
There are two cases:
1. $\sqrt{x^2 + y^2} - 1 \geq 0$ and $x + y - 1 \leq 0$ or
2. $\sqrt{x^2 + y^2} - 1 \leq 0$ and $x + y - 1 \geq 0$

They correspond to the regions below the line and outside the circle or above the line and inside the circle.

37. (D)
$$\vec{z} = \vec{y} - \vec{x} = \begin{pmatrix} -1 \\ -16 \end{pmatrix} - \begin{pmatrix} 6 \\ 8 \end{pmatrix} = \begin{pmatrix} -7 \\ -24 \end{pmatrix}$$
$$|\vec{z}| = \sqrt{49 + 576} = \sqrt{625} = 25$$

38. (D)
$$\cos^{-1}(\sin 150°) = \cos^{-1}\left(\frac{1}{2}\right) = 60°$$

39. (E)
The relative differences between the numbers in the array do not change in choices I, II and III.

40. (C)
A. (3, 2, 4)
B. (3, 2, -1)
C. (0, 2, -1)
AB=5 and BC=3
$$AC = \sqrt{(3-0)^2 + (2-2)^2 + (4--1)^2}$$
$$= \sqrt{9 + 25} = \sqrt{34} = 5.83$$

41. (B)
$$x = r\cos\theta = 4\cos\frac{5\pi}{6} = -3.464$$
$$y = r\sin\theta = 4\sin\frac{5\pi}{6} = 2$$

42. (E)
$$\frac{(2n+4)!}{(2n+2)!} = \frac{(2n+4)\cdot(2n+3)\cdot(2n+2)!}{(2n+2)!} = (2n+4)(2n+3)$$

43. (D)
The solid that results is a sphere with a radius of 5 $\Rightarrow$ Volume $= \frac{4}{3}\pi \cdot 5^3 = 523.60$

44. (B)
$$f(-1,14) = (-1)^3 - 1 = -2$$
$$f(1,-1) = 1 - 1 = 0$$
$$f(-1,1) - f(1,-1) = -2 - 0 = -2$$

45. (A)
Resulting function is
$$y = 3(x + 1 - 1)^2 - 5 + 5 \Rightarrow y = 3x^2$$
$$x = 1 \Rightarrow y = 3$$

46. (B)
$$h = 5r$$
$$2\pi rh = 3000$$
$$2\pi r \cdot 5r = 3000$$
$$\Rightarrow r^2 = 95.493 \Rightarrow r = 9.772 \text{ in}$$
$$r = 9.997 \text{in} = 0.81434 \text{ ft}$$
$$h = 4.0717 \text{ft}$$
$$\Rightarrow V = \pi r^2 h = \pi \cdot 0.81^2 \cdot 4.07 = 8.5$$

47. (B)
Indirect proof means proving the contrapositive of a statement instead if itself; that is proving $q' \Rightarrow p'$ instead of $p \Rightarrow q$. Therefore proof will start with the assumption given by q' in choice B.

48. (B)
$\beta = \{(l_1,l_2),(l_2,l_3),(l_4,l_5)\} \Rightarrow \beta$ has 3 elements.
Please note that the points on l_7 are not included in the relation.

49. (C)
$$f(x) = |3x|$$
$$f(-x) = |3(-x)| = |-3x| = |3x| = f(x)$$
Answer is C.

50. (E)
The distance from C to AB is the radius of the circle.
$$\tan\left(\frac{\hat{ACB}}{2}\right) = \tan 60° = \sqrt{3} = \frac{6\sqrt{3}}{R} \Rightarrow R = 6$$

Area of the shaded region is
$$\frac{12\sqrt{3} \cdot 6}{2} - \frac{\pi \cdot 6^2}{360°} \cdot 120° = 24.65$$

Model Test 15 – Solutions

1. (E)
A. for a=2 and b=3, ab will be even
B. for a=3 and b=2, 4a+b will be even
C. for a=2 and b=3, a+b+5 will be even
D. for a=3 and b=5, ab-1 will be even
E. 2(a+1)-2b+1=2(a+1-b)+1 is always odd regardless of a and b

2. (C)
$5 \cdot 4^2 = 80$

3. (B)
Two parallel lines are parallel if they have the same slope (I).
If both have a slope of 1, then their slopes are reciprocals of each other and they still do not intersect (II).

4. (B)
$x^2 + y^2 = 5$
$x^2 - y^2 = 3$
Add the equations above side by side
$2x^2 = 8 \Rightarrow x^2 = 4 \Rightarrow x = \pm 2$

5. (B)
$\frac{5}{6}x \neq 0 \Rightarrow x \neq 0$; therefore $\frac{5}{6} - x \neq \frac{5}{6}$

6. (A)
$\frac{2x}{y} - \frac{2x}{z} = \frac{2xz - 2xy}{yz}$

7. (B)
$2^{x-1} = \frac{1}{3} \Rightarrow 2^{x-1} - \frac{1}{3} = 0$ The graph of y= $2^{x-1} - \frac{1}{3}$
intersects the x axis when x = − 0.58.

8. (D)
$\log(\sin 20) + \log(\sin 20°) = (-0.039) + (-0.4659)$
$= -0.5055 \approx -0.506$

9. (B)
for y=-5, x=-7 and z=-9, x+y+z=-21

10. (C)
2k-1 is an odd integer. So, -4x must be negative and x must be positive.

11. (E)
$2@C = 2^C - C^2 = 0$
$C = -0.76 \text{ or } C = 2 \text{ or } C = 4$

12. (C)
II and III are correct only.

13. (A)
$y - z = 9 \Rightarrow 3t^2 + 4 - (t^3 - 1) = 9$
The graph of y = $3x^2 + 4 - (x^3 - 1) - 9$ intersects the x axis when x = -1.

14. (D)
I represents a function. II does not represent a function because if x is 1, y can be 1 or -1. III represents a function as it has only one element which is (0, 0).

15. (D)
$\sin(\pi - \theta) = \frac{x}{|AC|} \Rightarrow |AC| = \frac{x}{\sin(\pi - \theta)} = \frac{x}{\sin \theta}$

16. (B)
$f(g(2)) = f(2^3 + 2 + 1) = f(11) = \sqrt{3 \cdot 11 - 8} = 5$

17. (D)
5-x>0; x-1>0; and 5-x ≠ 1
1 < x < 5 and x ≠ 4

18. (B)
$f(1) = 2 \Rightarrow f^{-1}(2) = 1$; $f(2) = 3 \Rightarrow f^{-1}(3) = 2$
$f(3) = 4 \Rightarrow f^{-1}(4) = 3$; $f(4) = 1 \Rightarrow f^{-1}(1) = 4$

19. (A)
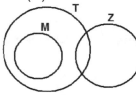
Let Z stand for Zandies, M for Mandies and T for Tandies. The suitable Venn diagram is as indicated in the figure.

20. (B)
$\sin(2\pi - \alpha) = -\sin(\alpha)$
$\cos(2\pi - \alpha) = \cos \alpha \Rightarrow x = 2\pi - \alpha$

21. (C)
$f(A) = A^2 - 9 \Rightarrow f(f(A)) = f(A^2 - 9) = (A^2 - 9)^2 - 9 = 0$
$\Rightarrow (A^2 - 9 - 3)(A^2 - 9 + 3) = 0 \Rightarrow A^2 = 12 \text{ or } A^2 = 6$
$\Rightarrow A = \mp 3.5 \text{ or } A = \mp 2.4$

22. (D)
$x^3 < x \Rightarrow x^3 - x < 0 \Rightarrow x(x - 1)(x + 1) < 0$
$\Rightarrow x < -1 \text{ or } 0 < x < 1$

23. (D)
I. $g(-x) = f^2(-x)$
II. $g(-x) = f(-x) + f(-(-x)) = f(-x) + f(x) = g(x)$
III. $g(-x) = f(-|-x|) = f(-|x|) = g(x)$

24. (B)

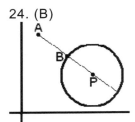

The point on circle P which is closest to point A is point B that is the intersection of AP with circle P.

Therefore

AB=AP-r= $\sqrt{(3-8)^2+(15-3)^2}-2=13-2=11$

25. (B)

$g(f(2,1))=g(\sqrt{3\cdot2-4\cdot1})=g(\sqrt{8})=3^{\sqrt{8}}\approx22.4$

26. (C)

$\left.\begin{array}{l}f(x)=ax^2+bx+c\\f(1)=4=a+b+c\\f(0)=3=c\\f(-1)=6=a-b+c\end{array}\right\}\begin{array}{l}a+b=1\\a-b=3\end{array}$

a=2 and b=-1

$y=2x^2-x+3$

$\Delta=(-1)^2-4\cdot2\cdot3<0$

The function is quadratic, it does not intersect the x axis and it opens upward. Therefore it's a positive function of x.

27. (C)

The point (3,-1) is the center and 10 is the radius of the circle that has the equation given by $(x-3)^2+(y+1)^2=10^2$.

28. (C)

BC=4=a where a is any of the base.

Base area= $6\cdot\dfrac{4^2\sqrt{3}}{4}=24\sqrt{3}$

Volume= $\dfrac{1}{3}\cdot24\sqrt{3}\cdot6=48\sqrt{3}$

29. (D)

$4t+12t+2=2\cdot(10t-1)\Rightarrow t=1$

$a_1=4t=4$

$a_2=10t-1=9$

d=9-4=5.

$a_{50}=a_1+49d=4+49\cdot5=249.$

30. (A)

P(x-2)=(x+1)Q(x)+R where R is the remainder when P(x-2) is divided by x+1. For x=-1, P(-1-2)=P(-3)=R.

31 (E)

$V=\dfrac{1}{2}\cdot\left(\pi\cdot4^2\cdot6+\dfrac{1}{3}\pi\cdot4^2\cdot4\right)\approx184$

32 (C)

In order to plot f(|x|), we omit the part of the graph of f(x) where x<0, then reflect and duplicate the part of f(x) where x>0. Therefore best choice for f(x) is C.

33. (E)

$(x+4\ge0\Rightarrow x\ge-4)$ and $(x-3\ne0\Rightarrow x\ne3)$

34. (C)

$\dfrac{\frac{2\pi}{5}}{2\pi}\cdot(\pi\cdot5^2-\pi\cdot3^2)=\dfrac{1}{5}\cdot16\pi=\dfrac{16\pi}{5}$

35. (B)

The distance between A(3,5) and the line 3x-4y+1=0

is: $\dfrac{|3\cdot3-4\cdot5+1|}{\sqrt{3^2+(-4)^2}}=\dfrac{10}{5}=2$ So,

AB= $2\cdot2=4$

36. (C)

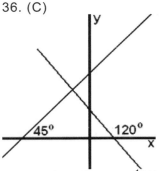

y = x + 3 makes $\tan^{-1}(1)$=45° with the positive x axis. y = $-\sqrt{3}$ x + 2 makes $\tan^{-1}(-\sqrt{3})$ = 120° with the positive x axis. The acute angle between the lines given is 120°-45°=75°, therefore the obtuse angle is 180°-75°=105°.

37. (A)

Increasing A would change amplitude only.

38. (A)

	Boys	Girls
Participated	10	7
Did not Participate	13	12

The probability is: $\dfrac{10+13+7}{23+19}=\dfrac{30}{42}=\dfrac{5}{7}$

39. (D)

Amplitude= $\dfrac{y_{max}-y_{min}}{2}=\dfrac{2-(-1)}{2}=1.5$

40. (D)

The positive zero's are:
$\{3, 5, 3+6, 5+6, 3+12, 5+12, 3+18, 5+18, 3+24, 5+24,...\} \Rightarrow \{3, 5, 9, 11, 15, 17, 21, 23, 27, 29,...\}$. The 10th positive zero is 29.

41. (A)

$$30 = A + B \cdot \sin\left(\frac{\pi}{2} - \frac{\pi \cdot 6}{6}\right) = A + B \cdot \sin\left(\frac{-\pi}{2}\right) = A - B$$

$$50 = A + B \cdot \sin\left(\frac{\pi}{2} - \frac{\pi \cdot 12}{6}\right) = A + B \cdot \sin\left(\frac{-3\pi}{2}\right) = A + B$$

$A - B = 30$
$A + B = 50$
Therefore A = 40 and B = 10.

So,

$$d(t) = 40 + 10 \cdot \sin\left(\frac{\pi}{2} - \frac{\pi t}{6}\right)$$

$$\Rightarrow 35 = 40 + 10 \cdot \sin\left(\frac{\pi}{2} - \frac{\pi t}{6}\right) \Rightarrow -\frac{1}{2} = \sin\left(\frac{\pi}{2} - \frac{\pi t}{6}\right)$$

$$-\frac{1}{2} = \cos\left(\frac{\pi t}{6}\right) \Rightarrow \frac{\pi t}{6} = \frac{2\pi}{3} \text{ or } \frac{\pi t}{6} = \frac{4\pi}{3}$$

$t = 4$ or $t = 8$

42. (B)

For the function $f(x) = 2x^2 + 12x + 3$

$h = -\dfrac{b}{2a} = -\dfrac{12}{2 \cdot 2} = -3$ where h is the abscissa of

the vertex. If we translate f(x), 3 units to the right, we get a function symmetric about the y-axis. So, k must be 3.

43. (D)

$\left(\dfrac{f}{g}\right)(x) = \dfrac{f(x)}{g(x)}$ has at least two points of

discontinuity. One of them is the point where g(x) = 0 and the other one is the point where g(x) is discontinuous itself (the corner point).

44. (E)

The roots are $x_1 = -2$, $x_2 = 3i$ and $x_3 = -3i$. So,
$y = (x + 2)(x + 3i)(x - 3i)$
$y = (x + 2)(x^2 + 9) = x^3 + 9x + 2x^2 + 18$
$\Rightarrow y = x^3 + 2x^2 + 9x + 18$

45. (B)

$$A\left(\overset{\Delta}{ABC}\right) = \frac{1}{2} \cdot 300 \cdot 150 \cdot \sin(90 - \theta) =$$

$$150^2 \cdot \cos\theta = 150^2 \cdot \frac{45}{75} = 13,500 \text{ square feet.}$$

46. (C)

$$\frac{-f(a)}{g(c)} - f(a) - g(c) = \frac{-b}{b} - b - b = -2b - 1$$

47. (D)

Standard deviation of the data set A is greater than that of B because in set B, the elements are more closely grouped about the mean.

48. (C)

I is not correct.
$$f(x, -y) = x^2 - x(-y) + (-y)^2$$
$$= x^2 + xy + y^2 \neq f(x, y)$$
II is not correct.
$$f(-x, y) = (-x)^2 - (-x)y + y^2$$
$$= x^2 + xy + y^2 \neq f(x, y)$$
III is correct.
$$f(-x, -y) = (-x)^2 - (-x)(-y) + (-y)^2$$
$$= x^2 - xy + y^2 = f(x, y)$$

49. (C)

Correct answer can be calculated by the matrix multiplication given in C.

50. (E)

$$x^2 - 4x + 4 - 4 - 4y^2 - 24y - 36 + 36 - 32 = 0$$

$$\Rightarrow (x - 2)^2 = (2y + 6)^2 \Rightarrow (x - 2) = \pm(2y + 6)$$

Which represents two lines that intersect at the same point.

Model Test 1 Answer Sheet

1.	Ⓐ	Ⓑ	Ⓒ	Ⓓ	Ⓔ	26.	Ⓐ	Ⓑ	Ⓒ	Ⓓ	Ⓔ
2.	Ⓐ	Ⓑ	Ⓒ	Ⓓ	Ⓔ	27.	Ⓐ	Ⓑ	Ⓒ	Ⓓ	Ⓔ
3.	Ⓐ	Ⓑ	Ⓒ	Ⓓ	Ⓔ	28.	Ⓐ	Ⓑ	Ⓒ	Ⓓ	Ⓔ
4.	Ⓐ	Ⓑ	Ⓒ	Ⓓ	Ⓔ	29.	Ⓐ	Ⓑ	Ⓒ	Ⓓ	Ⓔ
5.	Ⓐ	Ⓑ	Ⓒ	Ⓓ	Ⓔ	30.	Ⓐ	Ⓑ	Ⓒ	Ⓓ	Ⓔ
6.	Ⓐ	Ⓑ	Ⓒ	Ⓓ	Ⓔ	31.	Ⓐ	Ⓑ	Ⓒ	Ⓓ	Ⓔ
7.	Ⓐ	Ⓑ	Ⓒ	Ⓓ	Ⓔ	32.	Ⓐ	Ⓑ	Ⓒ	Ⓓ	Ⓔ
8.	Ⓐ	Ⓑ	Ⓒ	Ⓓ	Ⓔ	33.	Ⓐ	Ⓑ	Ⓒ	Ⓓ	Ⓔ
9.	Ⓐ	Ⓑ	Ⓒ	Ⓓ	Ⓔ	34.	Ⓐ	Ⓑ	Ⓒ	Ⓓ	Ⓔ
10.	Ⓐ	Ⓑ	Ⓒ	Ⓓ	Ⓔ	35.	Ⓐ	Ⓑ	Ⓒ	Ⓓ	Ⓔ
11.	Ⓐ	Ⓑ	Ⓒ	Ⓓ	Ⓔ	36.	Ⓐ	Ⓑ	Ⓒ	Ⓓ	Ⓔ
12.	Ⓐ	Ⓑ	Ⓒ	Ⓓ	Ⓔ	37.	Ⓐ	Ⓑ	Ⓒ	Ⓓ	Ⓔ
13.	Ⓐ	Ⓑ	Ⓒ	Ⓓ	Ⓔ	38.	Ⓐ	Ⓑ	Ⓒ	Ⓓ	Ⓔ
14.	Ⓐ	Ⓑ	Ⓒ	Ⓓ	Ⓔ	39.	Ⓐ	Ⓑ	Ⓒ	Ⓓ	Ⓔ
15.	Ⓐ	Ⓑ	Ⓒ	Ⓓ	Ⓔ	40.	Ⓐ	Ⓑ	Ⓒ	Ⓓ	Ⓔ
16.	Ⓐ	Ⓑ	Ⓒ	Ⓓ	Ⓔ	41.	Ⓐ	Ⓑ	Ⓒ	Ⓓ	Ⓔ
17.	Ⓐ	Ⓑ	Ⓒ	Ⓓ	Ⓔ	42.	Ⓐ	Ⓑ	Ⓒ	Ⓓ	Ⓔ
18.	Ⓐ	Ⓑ	Ⓒ	Ⓓ	Ⓔ	43.	Ⓐ	Ⓑ	Ⓒ	Ⓓ	Ⓔ
19.	Ⓐ	Ⓑ	Ⓒ	Ⓓ	Ⓔ	44.	Ⓐ	Ⓑ	Ⓒ	Ⓓ	Ⓔ
20.	Ⓐ	Ⓑ	Ⓒ	Ⓓ	Ⓔ	45.	Ⓐ	Ⓑ	Ⓒ	Ⓓ	Ⓔ
21.	Ⓐ	Ⓑ	Ⓒ	Ⓓ	Ⓔ	46.	Ⓐ	Ⓑ	Ⓒ	Ⓓ	Ⓔ
22.	Ⓐ	Ⓑ	Ⓒ	Ⓓ	Ⓔ	47.	Ⓐ	Ⓑ	Ⓒ	Ⓓ	Ⓔ
23.	Ⓐ	Ⓑ	Ⓒ	Ⓓ	Ⓔ	48.	Ⓐ	Ⓑ	Ⓒ	Ⓓ	Ⓔ
24.	Ⓐ	Ⓑ	Ⓒ	Ⓓ	Ⓔ	49.	Ⓐ	Ⓑ	Ⓒ	Ⓓ	Ⓔ
25.	Ⓐ	Ⓑ	Ⓒ	Ⓓ	Ⓔ	50.	Ⓐ	Ⓑ	Ⓒ	Ⓓ	Ⓔ

Model Test 2 Answer Sheet

1.	Ⓐ	Ⓑ	Ⓒ	Ⓓ	Ⓔ	26.	Ⓐ	Ⓑ	Ⓒ	Ⓓ	Ⓔ
2.	Ⓐ	Ⓑ	Ⓒ	Ⓓ	Ⓔ	27.	Ⓐ	Ⓑ	Ⓒ	Ⓓ	Ⓔ
3.	Ⓐ	Ⓑ	Ⓒ	Ⓓ	Ⓔ	28.	Ⓐ	Ⓑ	Ⓒ	Ⓓ	Ⓔ
4.	Ⓐ	Ⓑ	Ⓒ	Ⓓ	Ⓔ	29.	Ⓐ	Ⓑ	Ⓒ	Ⓓ	Ⓔ
5.	Ⓐ	Ⓑ	Ⓒ	Ⓓ	Ⓔ	30.	Ⓐ	Ⓑ	Ⓒ	Ⓓ	Ⓔ
6.	Ⓐ	Ⓑ	Ⓒ	Ⓓ	Ⓔ	31.	Ⓐ	Ⓑ	Ⓒ	Ⓓ	Ⓔ
7.	Ⓐ	Ⓑ	Ⓒ	Ⓓ	Ⓔ	32.	Ⓐ	Ⓑ	Ⓒ	Ⓓ	Ⓔ
8.	Ⓐ	Ⓑ	Ⓒ	Ⓓ	Ⓔ	33.	Ⓐ	Ⓑ	Ⓒ	Ⓓ	Ⓔ
9.	Ⓐ	Ⓑ	Ⓒ	Ⓓ	Ⓔ	34.	Ⓐ	Ⓑ	Ⓒ	Ⓓ	Ⓔ
10.	Ⓐ	Ⓑ	Ⓒ	Ⓓ	Ⓔ	35.	Ⓐ	Ⓑ	Ⓒ	Ⓓ	Ⓔ
11.	Ⓐ	Ⓑ	Ⓒ	Ⓓ	Ⓔ	36.	Ⓐ	Ⓑ	Ⓒ	Ⓓ	Ⓔ
12.	Ⓐ	Ⓑ	Ⓒ	Ⓓ	Ⓔ	37.	Ⓐ	Ⓑ	Ⓒ	Ⓓ	Ⓔ
13.	Ⓐ	Ⓑ	Ⓒ	Ⓓ	Ⓔ	38.	Ⓐ	Ⓑ	Ⓒ	Ⓓ	Ⓔ
14.	Ⓐ	Ⓑ	Ⓒ	Ⓓ	Ⓔ	39.	Ⓐ	Ⓑ	Ⓒ	Ⓓ	Ⓔ
15.	Ⓐ	Ⓑ	Ⓒ	Ⓓ	Ⓔ	40.	Ⓐ	Ⓑ	Ⓒ	Ⓓ	Ⓔ
16.	Ⓐ	Ⓑ	Ⓒ	Ⓓ	Ⓔ	41.	Ⓐ	Ⓑ	Ⓒ	Ⓓ	Ⓔ
17.	Ⓐ	Ⓑ	Ⓒ	Ⓓ	Ⓔ	42.	Ⓐ	Ⓑ	Ⓒ	Ⓓ	Ⓔ
18.	Ⓐ	Ⓑ	Ⓒ	Ⓓ	Ⓔ	43.	Ⓐ	Ⓑ	Ⓒ	Ⓓ	Ⓔ
19.	Ⓐ	Ⓑ	Ⓒ	Ⓓ	Ⓔ	44.	Ⓐ	Ⓑ	Ⓒ	Ⓓ	Ⓔ
20.	Ⓐ	Ⓑ	Ⓒ	Ⓓ	Ⓔ	45.	Ⓐ	Ⓑ	Ⓒ	Ⓓ	Ⓔ
21.	Ⓐ	Ⓑ	Ⓒ	Ⓓ	Ⓔ	46.	Ⓐ	Ⓑ	Ⓒ	Ⓓ	Ⓔ
22.	Ⓐ	Ⓑ	Ⓒ	Ⓓ	Ⓔ	47.	Ⓐ	Ⓑ	Ⓒ	Ⓓ	Ⓔ
23.	Ⓐ	Ⓑ	Ⓒ	Ⓓ	Ⓔ	48.	Ⓐ	Ⓑ	Ⓒ	Ⓓ	Ⓔ
24.	Ⓐ	Ⓑ	Ⓒ	Ⓓ	Ⓔ	49.	Ⓐ	Ⓑ	Ⓒ	Ⓓ	Ⓔ
25.	Ⓐ	Ⓑ	Ⓒ	Ⓓ	Ⓔ	50.	Ⓐ	Ⓑ	Ⓒ	Ⓓ	Ⓔ

Model Test 3 Answer Sheet

#							#					
1.	Ⓐ	Ⓑ	Ⓒ	Ⓓ	Ⓔ		26.	Ⓐ	Ⓑ	Ⓒ	Ⓓ	Ⓔ
2.	Ⓐ	Ⓑ	Ⓒ	Ⓓ	Ⓔ		27.	Ⓐ	Ⓑ	Ⓒ	Ⓓ	Ⓔ
3.	Ⓐ	Ⓑ	Ⓒ	Ⓓ	Ⓔ		28.	Ⓐ	Ⓑ	Ⓒ	Ⓓ	Ⓔ
4.	Ⓐ	Ⓑ	Ⓒ	Ⓓ	Ⓔ		29.	Ⓐ	Ⓑ	Ⓒ	Ⓓ	Ⓔ
5.	Ⓐ	Ⓑ	Ⓒ	Ⓓ	Ⓔ		30.	Ⓐ	Ⓑ	Ⓒ	Ⓓ	Ⓔ
6.	Ⓐ	Ⓑ	Ⓒ	Ⓓ	Ⓔ		31.	Ⓐ	Ⓑ	Ⓒ	Ⓓ	Ⓔ
7.	Ⓐ	Ⓑ	Ⓒ	Ⓓ	Ⓔ		32.	Ⓐ	Ⓑ	Ⓒ	Ⓓ	Ⓔ
8.	Ⓐ	Ⓑ	Ⓒ	Ⓓ	Ⓔ		33.	Ⓐ	Ⓑ	Ⓒ	Ⓓ	Ⓔ
9.	Ⓐ	Ⓑ	Ⓒ	Ⓓ	Ⓔ		34.	Ⓐ	Ⓑ	Ⓒ	Ⓓ	Ⓔ
10.	Ⓐ	Ⓑ	Ⓒ	Ⓓ	Ⓔ		35.	Ⓐ	Ⓑ	Ⓒ	Ⓓ	Ⓔ
11.	Ⓐ	Ⓑ	Ⓒ	Ⓓ	Ⓔ		36.	Ⓐ	Ⓑ	Ⓒ	Ⓓ	Ⓔ
12.	Ⓐ	Ⓑ	Ⓒ	Ⓓ	Ⓔ		37.	Ⓐ	Ⓑ	Ⓒ	Ⓓ	Ⓔ
13.	Ⓐ	Ⓑ	Ⓒ	Ⓓ	Ⓔ		38.	Ⓐ	Ⓑ	Ⓒ	Ⓓ	Ⓔ
14.	Ⓐ	Ⓑ	Ⓒ	Ⓓ	Ⓔ		39.	Ⓐ	Ⓑ	Ⓒ	Ⓓ	Ⓔ
15.	Ⓐ	Ⓑ	Ⓒ	Ⓓ	Ⓔ		40.	Ⓐ	Ⓑ	Ⓒ	Ⓓ	Ⓔ
16.	Ⓐ	Ⓑ	Ⓒ	Ⓓ	Ⓔ		41.	Ⓐ	Ⓑ	Ⓒ	Ⓓ	Ⓔ
17.	Ⓐ	Ⓑ	Ⓒ	Ⓓ	Ⓔ		42.	Ⓐ	Ⓑ	Ⓒ	Ⓓ	Ⓔ
18.	Ⓐ	Ⓑ	Ⓒ	Ⓓ	Ⓔ		43.	Ⓐ	Ⓑ	Ⓒ	Ⓓ	Ⓔ
19.	Ⓐ	Ⓑ	Ⓒ	Ⓓ	Ⓔ		44.	Ⓐ	Ⓑ	Ⓒ	Ⓓ	Ⓔ
20.	Ⓐ	Ⓑ	Ⓒ	Ⓓ	Ⓔ		45.	Ⓐ	Ⓑ	Ⓒ	Ⓓ	Ⓔ
21.	Ⓐ	Ⓑ	Ⓒ	Ⓓ	Ⓔ		46.	Ⓐ	Ⓑ	Ⓒ	Ⓓ	Ⓔ
22.	Ⓐ	Ⓑ	Ⓒ	Ⓓ	Ⓔ		47.	Ⓐ	Ⓑ	Ⓒ	Ⓓ	Ⓔ
23.	Ⓐ	Ⓑ	Ⓒ	Ⓓ	Ⓔ		48.	Ⓐ	Ⓑ	Ⓒ	Ⓓ	Ⓔ
24.	Ⓐ	Ⓑ	Ⓒ	Ⓓ	Ⓔ		49.	Ⓐ	Ⓑ	Ⓒ	Ⓓ	Ⓔ
25.	Ⓐ	Ⓑ	Ⓒ	Ⓓ	Ⓔ		50.	Ⓐ	Ⓑ	Ⓒ	Ⓓ	Ⓔ

Model Test 4 Answer Sheet

#							#					
1.	Ⓐ	Ⓑ	Ⓒ	Ⓓ	Ⓔ		26.	Ⓐ	Ⓑ	Ⓒ	Ⓓ	Ⓔ
2.	Ⓐ	Ⓑ	Ⓒ	Ⓓ	Ⓔ		27.	Ⓐ	Ⓑ	Ⓒ	Ⓓ	Ⓔ
3.	Ⓐ	Ⓑ	Ⓒ	Ⓓ	Ⓔ		28.	Ⓐ	Ⓑ	Ⓒ	Ⓓ	Ⓔ
4.	Ⓐ	Ⓑ	Ⓒ	Ⓓ	Ⓔ		29.	Ⓐ	Ⓑ	Ⓒ	Ⓓ	Ⓔ
5.	Ⓐ	Ⓑ	Ⓒ	Ⓓ	Ⓔ		30.	Ⓐ	Ⓑ	Ⓒ	Ⓓ	Ⓔ
6.	Ⓐ	Ⓑ	Ⓒ	Ⓓ	Ⓔ		31.	Ⓐ	Ⓑ	Ⓒ	Ⓓ	Ⓔ
7.	Ⓐ	Ⓑ	Ⓒ	Ⓓ	Ⓔ		32.	Ⓐ	Ⓑ	Ⓒ	Ⓓ	Ⓔ
8.	Ⓐ	Ⓑ	Ⓒ	Ⓓ	Ⓔ		33.	Ⓐ	Ⓑ	Ⓒ	Ⓓ	Ⓔ
9.	Ⓐ	Ⓑ	Ⓒ	Ⓓ	Ⓔ		34.	Ⓐ	Ⓑ	Ⓒ	Ⓓ	Ⓔ
10.	Ⓐ	Ⓑ	Ⓒ	Ⓓ	Ⓔ		35.	Ⓐ	Ⓑ	Ⓒ	Ⓓ	Ⓔ
11.	Ⓐ	Ⓑ	Ⓒ	Ⓓ	Ⓔ		36.	Ⓐ	Ⓑ	Ⓒ	Ⓓ	Ⓔ
12.	Ⓐ	Ⓑ	Ⓒ	Ⓓ	Ⓔ		37.	Ⓐ	Ⓑ	Ⓒ	Ⓓ	Ⓔ
13.	Ⓐ	Ⓑ	Ⓒ	Ⓓ	Ⓔ		38.	Ⓐ	Ⓑ	Ⓒ	Ⓓ	Ⓔ
14.	Ⓐ	Ⓑ	Ⓒ	Ⓓ	Ⓔ		39.	Ⓐ	Ⓑ	Ⓒ	Ⓓ	Ⓔ
15.	Ⓐ	Ⓑ	Ⓒ	Ⓓ	Ⓔ		40.	Ⓐ	Ⓑ	Ⓒ	Ⓓ	Ⓔ
16.	Ⓐ	Ⓑ	Ⓒ	Ⓓ	Ⓔ		41.	Ⓐ	Ⓑ	Ⓒ	Ⓓ	Ⓔ
17.	Ⓐ	Ⓑ	Ⓒ	Ⓓ	Ⓔ		42.	Ⓐ	Ⓑ	Ⓒ	Ⓓ	Ⓔ
18.	Ⓐ	Ⓑ	Ⓒ	Ⓓ	Ⓔ		43.	Ⓐ	Ⓑ	Ⓒ	Ⓓ	Ⓔ
19.	Ⓐ	Ⓑ	Ⓒ	Ⓓ	Ⓔ		44.	Ⓐ	Ⓑ	Ⓒ	Ⓓ	Ⓔ
20.	Ⓐ	Ⓑ	Ⓒ	Ⓓ	Ⓔ		45.	Ⓐ	Ⓑ	Ⓒ	Ⓓ	Ⓔ
21.	Ⓐ	Ⓑ	Ⓒ	Ⓓ	Ⓔ		46.	Ⓐ	Ⓑ	Ⓒ	Ⓓ	Ⓔ
22.	Ⓐ	Ⓑ	Ⓒ	Ⓓ	Ⓔ		47.	Ⓐ	Ⓑ	Ⓒ	Ⓓ	Ⓔ
23.	Ⓐ	Ⓑ	Ⓒ	Ⓓ	Ⓔ		48.	Ⓐ	Ⓑ	Ⓒ	Ⓓ	Ⓔ
24.	Ⓐ	Ⓑ	Ⓒ	Ⓓ	Ⓔ		49.	Ⓐ	Ⓑ	Ⓒ	Ⓓ	Ⓔ
25.	Ⓐ	Ⓑ	Ⓒ	Ⓓ	Ⓔ		50.	Ⓐ	Ⓑ	Ⓒ	Ⓓ	Ⓔ

Model Test 5 Answer Sheet

#	A	B	C	D	E		#	A	B	C	D	E
1.	Ⓐ	Ⓑ	Ⓒ	Ⓓ	Ⓔ		26.	Ⓐ	Ⓑ	Ⓒ	Ⓓ	Ⓔ
2.	Ⓐ	Ⓑ	Ⓒ	Ⓓ	Ⓔ		27.	Ⓐ	Ⓑ	Ⓒ	Ⓓ	Ⓔ
3.	Ⓐ	Ⓑ	Ⓒ	Ⓓ	Ⓔ		28.	Ⓐ	Ⓑ	Ⓒ	Ⓓ	Ⓔ
4.	Ⓐ	Ⓑ	Ⓒ	Ⓓ	Ⓔ		29.	Ⓐ	Ⓑ	Ⓒ	Ⓓ	Ⓔ
5.	Ⓐ	Ⓑ	Ⓒ	Ⓓ	Ⓔ		30.	Ⓐ	Ⓑ	Ⓒ	Ⓓ	Ⓔ
6.	Ⓐ	Ⓑ	Ⓒ	Ⓓ	Ⓔ		31.	Ⓐ	Ⓑ	Ⓒ	Ⓓ	Ⓔ
7.	Ⓐ	Ⓑ	Ⓒ	Ⓓ	Ⓔ		32.	Ⓐ	Ⓑ	Ⓒ	Ⓓ	Ⓔ
8.	Ⓐ	Ⓑ	Ⓒ	Ⓓ	Ⓔ		33.	Ⓐ	Ⓑ	Ⓒ	Ⓓ	Ⓔ
9.	Ⓐ	Ⓑ	Ⓒ	Ⓓ	Ⓔ		34.	Ⓐ	Ⓑ	Ⓒ	Ⓓ	Ⓔ
10.	Ⓐ	Ⓑ	Ⓒ	Ⓓ	Ⓔ		35.	Ⓐ	Ⓑ	Ⓒ	Ⓓ	Ⓔ
11.	Ⓐ	Ⓑ	Ⓒ	Ⓓ	Ⓔ		36.	Ⓐ	Ⓑ	Ⓒ	Ⓓ	Ⓔ
12.	Ⓐ	Ⓑ	Ⓒ	Ⓓ	Ⓔ		37.	Ⓐ	Ⓑ	Ⓒ	Ⓓ	Ⓔ
13.	Ⓐ	Ⓑ	Ⓒ	Ⓓ	Ⓔ		38.	Ⓐ	Ⓑ	Ⓒ	Ⓓ	Ⓔ
14.	Ⓐ	Ⓑ	Ⓒ	Ⓓ	Ⓔ		39.	Ⓐ	Ⓑ	Ⓒ	Ⓓ	Ⓔ
15.	Ⓐ	Ⓑ	Ⓒ	Ⓓ	Ⓔ		40.	Ⓐ	Ⓑ	Ⓒ	Ⓓ	Ⓔ
16.	Ⓐ	Ⓑ	Ⓒ	Ⓓ	Ⓔ		41.	Ⓐ	Ⓑ	Ⓒ	Ⓓ	Ⓔ
17.	Ⓐ	Ⓑ	Ⓒ	Ⓓ	Ⓔ		42.	Ⓐ	Ⓑ	Ⓒ	Ⓓ	Ⓔ
18.	Ⓐ	Ⓑ	Ⓒ	Ⓓ	Ⓔ		43.	Ⓐ	Ⓑ	Ⓒ	Ⓓ	Ⓔ
19.	Ⓐ	Ⓑ	Ⓒ	Ⓓ	Ⓔ		44.	Ⓐ	Ⓑ	Ⓒ	Ⓓ	Ⓔ
20.	Ⓐ	Ⓑ	Ⓒ	Ⓓ	Ⓔ		45.	Ⓐ	Ⓑ	Ⓒ	Ⓓ	Ⓔ
21.	Ⓐ	Ⓑ	Ⓒ	Ⓓ	Ⓔ		46.	Ⓐ	Ⓑ	Ⓒ	Ⓓ	Ⓔ
22.	Ⓐ	Ⓑ	Ⓒ	Ⓓ	Ⓔ		47.	Ⓐ	Ⓑ	Ⓒ	Ⓓ	Ⓔ
23.	Ⓐ	Ⓑ	Ⓒ	Ⓓ	Ⓔ		48.	Ⓐ	Ⓑ	Ⓒ	Ⓓ	Ⓔ
24.	Ⓐ	Ⓑ	Ⓒ	Ⓓ	Ⓔ		49.	Ⓐ	Ⓑ	Ⓒ	Ⓓ	Ⓔ
25.	Ⓐ	Ⓑ	Ⓒ	Ⓓ	Ⓔ		50.	Ⓐ	Ⓑ	Ⓒ	Ⓓ	Ⓔ

Model Test 6 Answer Sheet

#	A	B	C	D	E		#	A	B	C	D	E
1.	Ⓐ	Ⓑ	Ⓒ	Ⓓ	Ⓔ		26.	Ⓐ	Ⓑ	Ⓒ	Ⓓ	Ⓔ
2.	Ⓐ	Ⓑ	Ⓒ	Ⓓ	Ⓔ		27.	Ⓐ	Ⓑ	Ⓒ	Ⓓ	Ⓔ
3.	Ⓐ	Ⓑ	Ⓒ	Ⓓ	Ⓔ		28.	Ⓐ	Ⓑ	Ⓒ	Ⓓ	Ⓔ
4.	Ⓐ	Ⓑ	Ⓒ	Ⓓ	Ⓔ		29.	Ⓐ	Ⓑ	Ⓒ	Ⓓ	Ⓔ
5.	Ⓐ	Ⓑ	Ⓒ	Ⓓ	Ⓔ		30.	Ⓐ	Ⓑ	Ⓒ	Ⓓ	Ⓔ
6.	Ⓐ	Ⓑ	Ⓒ	Ⓓ	Ⓔ		31.	Ⓐ	Ⓑ	Ⓒ	Ⓓ	Ⓔ
7.	Ⓐ	Ⓑ	Ⓒ	Ⓓ	Ⓔ		32.	Ⓐ	Ⓑ	Ⓒ	Ⓓ	Ⓔ
8.	Ⓐ	Ⓑ	Ⓒ	Ⓓ	Ⓔ		33.	Ⓐ	Ⓑ	Ⓒ	Ⓓ	Ⓔ
9.	Ⓐ	Ⓑ	Ⓒ	Ⓓ	Ⓔ		34.	Ⓐ	Ⓑ	Ⓒ	Ⓓ	Ⓔ
10.	Ⓐ	Ⓑ	Ⓒ	Ⓓ	Ⓔ		35.	Ⓐ	Ⓑ	Ⓒ	Ⓓ	Ⓔ
11.	Ⓐ	Ⓑ	Ⓒ	Ⓓ	Ⓔ		36.	Ⓐ	Ⓑ	Ⓒ	Ⓓ	Ⓔ
12.	Ⓐ	Ⓑ	Ⓒ	Ⓓ	Ⓔ		37.	Ⓐ	Ⓑ	Ⓒ	Ⓓ	Ⓔ
13.	Ⓐ	Ⓑ	Ⓒ	Ⓓ	Ⓔ		38.	Ⓐ	Ⓑ	Ⓒ	Ⓓ	Ⓔ
14.	Ⓐ	Ⓑ	Ⓒ	Ⓓ	Ⓔ		39.	Ⓐ	Ⓑ	Ⓒ	Ⓓ	Ⓔ
15.	Ⓐ	Ⓑ	Ⓒ	Ⓓ	Ⓔ		40.	Ⓐ	Ⓑ	Ⓒ	Ⓓ	Ⓔ
16.	Ⓐ	Ⓑ	Ⓒ	Ⓓ	Ⓔ		41.	Ⓐ	Ⓑ	Ⓒ	Ⓓ	Ⓔ
17.	Ⓐ	Ⓑ	Ⓒ	Ⓓ	Ⓔ		42.	Ⓐ	Ⓑ	Ⓒ	Ⓓ	Ⓔ
18.	Ⓐ	Ⓑ	Ⓒ	Ⓓ	Ⓔ		43.	Ⓐ	Ⓑ	Ⓒ	Ⓓ	Ⓔ
19.	Ⓐ	Ⓑ	Ⓒ	Ⓓ	Ⓔ		44.	Ⓐ	Ⓑ	Ⓒ	Ⓓ	Ⓔ
20.	Ⓐ	Ⓑ	Ⓒ	Ⓓ	Ⓔ		45.	Ⓐ	Ⓑ	Ⓒ	Ⓓ	Ⓔ
21.	Ⓐ	Ⓑ	Ⓒ	Ⓓ	Ⓔ		46.	Ⓐ	Ⓑ	Ⓒ	Ⓓ	Ⓔ
22.	Ⓐ	Ⓑ	Ⓒ	Ⓓ	Ⓔ		47.	Ⓐ	Ⓑ	Ⓒ	Ⓓ	Ⓔ
23.	Ⓐ	Ⓑ	Ⓒ	Ⓓ	Ⓔ		48.	Ⓐ	Ⓑ	Ⓒ	Ⓓ	Ⓔ
24.	Ⓐ	Ⓑ	Ⓒ	Ⓓ	Ⓔ		49.	Ⓐ	Ⓑ	Ⓒ	Ⓓ	Ⓔ
25.	Ⓐ	Ⓑ	Ⓒ	Ⓓ	Ⓔ		50.	Ⓐ	Ⓑ	Ⓒ	Ⓓ	Ⓔ

Model Test 7 Answer Sheet

1.	Ⓐ Ⓑ Ⓒ Ⓓ Ⓔ		26.	Ⓐ Ⓑ Ⓒ Ⓓ Ⓔ
2.	Ⓐ Ⓑ Ⓒ Ⓓ Ⓔ		27.	Ⓐ Ⓑ Ⓒ Ⓓ Ⓔ
3.	Ⓐ Ⓑ Ⓒ Ⓓ Ⓔ		28.	Ⓐ Ⓑ Ⓒ Ⓓ Ⓔ
4.	Ⓐ Ⓑ Ⓒ Ⓓ Ⓔ		29.	Ⓐ Ⓑ Ⓒ Ⓓ Ⓔ
5.	Ⓐ Ⓑ Ⓒ Ⓓ Ⓔ		30.	Ⓐ Ⓑ Ⓒ Ⓓ Ⓔ
6.	Ⓐ Ⓑ Ⓒ Ⓓ Ⓔ		31.	Ⓐ Ⓑ Ⓒ Ⓓ Ⓔ
7.	Ⓐ Ⓑ Ⓒ Ⓓ Ⓔ		32.	Ⓐ Ⓑ Ⓒ Ⓓ Ⓔ
8.	Ⓐ Ⓑ Ⓒ Ⓓ Ⓔ		33.	Ⓐ Ⓑ Ⓒ Ⓓ Ⓔ
9.	Ⓐ Ⓑ Ⓒ Ⓓ Ⓔ		34.	Ⓐ Ⓑ Ⓒ Ⓓ Ⓔ
10.	Ⓐ Ⓑ Ⓒ Ⓓ Ⓔ		35.	Ⓐ Ⓑ Ⓒ Ⓓ Ⓔ
11.	Ⓐ Ⓑ Ⓒ Ⓓ Ⓔ		36.	Ⓐ Ⓑ Ⓒ Ⓓ Ⓔ
12.	Ⓐ Ⓑ Ⓒ Ⓓ Ⓔ		37.	Ⓐ Ⓑ Ⓒ Ⓓ Ⓔ
13.	Ⓐ Ⓑ Ⓒ Ⓓ Ⓔ		38.	Ⓐ Ⓑ Ⓒ Ⓓ Ⓔ
14.	Ⓐ Ⓑ Ⓒ Ⓓ Ⓔ		39.	Ⓐ Ⓑ Ⓒ Ⓓ Ⓔ
15.	Ⓐ Ⓑ Ⓒ Ⓓ Ⓔ		40.	Ⓐ Ⓑ Ⓒ Ⓓ Ⓔ
16.	Ⓐ Ⓑ Ⓒ Ⓓ Ⓔ		41.	Ⓐ Ⓑ Ⓒ Ⓓ Ⓔ
17.	Ⓐ Ⓑ Ⓒ Ⓓ Ⓔ		42.	Ⓐ Ⓑ Ⓒ Ⓓ Ⓔ
18.	Ⓐ Ⓑ Ⓒ Ⓓ Ⓔ		43.	Ⓐ Ⓑ Ⓒ Ⓓ Ⓔ
19.	Ⓐ Ⓑ Ⓒ Ⓓ Ⓔ		44.	Ⓐ Ⓑ Ⓒ Ⓓ Ⓔ
20.	Ⓐ Ⓑ Ⓒ Ⓓ Ⓔ		45.	Ⓐ Ⓑ Ⓒ Ⓓ Ⓔ
21.	Ⓐ Ⓑ Ⓒ Ⓓ Ⓔ		46.	Ⓐ Ⓑ Ⓒ Ⓓ Ⓔ
22.	Ⓐ Ⓑ Ⓒ Ⓓ Ⓔ		47.	Ⓐ Ⓑ Ⓒ Ⓓ Ⓔ
23.	Ⓐ Ⓑ Ⓒ Ⓓ Ⓔ		48.	Ⓐ Ⓑ Ⓒ Ⓓ Ⓔ
24.	Ⓐ Ⓑ Ⓒ Ⓓ Ⓔ		49.	Ⓐ Ⓑ Ⓒ Ⓓ Ⓔ
25.	Ⓐ Ⓑ Ⓒ Ⓓ Ⓔ		50.	Ⓐ Ⓑ Ⓒ Ⓓ Ⓔ

Model Test 8 Answer Sheet

1.	Ⓐ Ⓑ Ⓒ Ⓓ Ⓔ		26.	Ⓐ Ⓑ Ⓒ Ⓓ Ⓔ
2.	Ⓐ Ⓑ Ⓒ Ⓓ Ⓔ		27.	Ⓐ Ⓑ Ⓒ Ⓓ Ⓔ
3.	Ⓐ Ⓑ Ⓒ Ⓓ Ⓔ		28.	Ⓐ Ⓑ Ⓒ Ⓓ Ⓔ
4.	Ⓐ Ⓑ Ⓒ Ⓓ Ⓔ		29.	Ⓐ Ⓑ Ⓒ Ⓓ Ⓔ
5.	Ⓐ Ⓑ Ⓒ Ⓓ Ⓔ		30.	Ⓐ Ⓑ Ⓒ Ⓓ Ⓔ
6.	Ⓐ Ⓑ Ⓒ Ⓓ Ⓔ		31.	Ⓐ Ⓑ Ⓒ Ⓓ Ⓔ
7.	Ⓐ Ⓑ Ⓒ Ⓓ Ⓔ		32.	Ⓐ Ⓑ Ⓒ Ⓓ Ⓔ
8.	Ⓐ Ⓑ Ⓒ Ⓓ Ⓔ		33.	Ⓐ Ⓑ Ⓒ Ⓓ Ⓔ
9.	Ⓐ Ⓑ Ⓒ Ⓓ Ⓔ		34.	Ⓐ Ⓑ Ⓒ Ⓓ Ⓔ
10.	Ⓐ Ⓑ Ⓒ Ⓓ Ⓔ		35.	Ⓐ Ⓑ Ⓒ Ⓓ Ⓔ
11.	Ⓐ Ⓑ Ⓒ Ⓓ Ⓔ		36.	Ⓐ Ⓑ Ⓒ Ⓓ Ⓔ
12.	Ⓐ Ⓑ Ⓒ Ⓓ Ⓔ		37.	Ⓐ Ⓑ Ⓒ Ⓓ Ⓔ
13.	Ⓐ Ⓑ Ⓒ Ⓓ Ⓔ		38.	Ⓐ Ⓑ Ⓒ Ⓓ Ⓔ
14.	Ⓐ Ⓑ Ⓒ Ⓓ Ⓔ		39.	Ⓐ Ⓑ Ⓒ Ⓓ Ⓔ
15.	Ⓐ Ⓑ Ⓒ Ⓓ Ⓔ		40.	Ⓐ Ⓑ Ⓒ Ⓓ Ⓔ
16.	Ⓐ Ⓑ Ⓒ Ⓓ Ⓔ		41.	Ⓐ Ⓑ Ⓒ Ⓓ Ⓔ
17.	Ⓐ Ⓑ Ⓒ Ⓓ Ⓔ		42.	Ⓐ Ⓑ Ⓒ Ⓓ Ⓔ
18.	Ⓐ Ⓑ Ⓒ Ⓓ Ⓔ		43.	Ⓐ Ⓑ Ⓒ Ⓓ Ⓔ
19.	Ⓐ Ⓑ Ⓒ Ⓓ Ⓔ		44.	Ⓐ Ⓑ Ⓒ Ⓓ Ⓔ
20.	Ⓐ Ⓑ Ⓒ Ⓓ Ⓔ		45.	Ⓐ Ⓑ Ⓒ Ⓓ Ⓔ
21.	Ⓐ Ⓑ Ⓒ Ⓓ Ⓔ		46.	Ⓐ Ⓑ Ⓒ Ⓓ Ⓔ
22.	Ⓐ Ⓑ Ⓒ Ⓓ Ⓔ		47.	Ⓐ Ⓑ Ⓒ Ⓓ Ⓔ
23.	Ⓐ Ⓑ Ⓒ Ⓓ Ⓔ		48.	Ⓐ Ⓑ Ⓒ Ⓓ Ⓔ
24.	Ⓐ Ⓑ Ⓒ Ⓓ Ⓔ		49.	Ⓐ Ⓑ Ⓒ Ⓓ Ⓔ
25.	Ⓐ Ⓑ Ⓒ Ⓓ Ⓔ		50.	Ⓐ Ⓑ Ⓒ Ⓓ Ⓔ

Model Test 9 Answer Sheet

#						#					
1.	Ⓐ	Ⓑ	Ⓒ	Ⓓ	Ⓔ	26.	Ⓐ	Ⓑ	Ⓒ	Ⓓ	Ⓔ
2.	Ⓐ	Ⓑ	Ⓒ	Ⓓ	Ⓔ	27.	Ⓐ	Ⓑ	Ⓒ	Ⓓ	Ⓔ
3.	Ⓐ	Ⓑ	Ⓒ	Ⓓ	Ⓔ	28.	Ⓐ	Ⓑ	Ⓒ	Ⓓ	Ⓔ
4.	Ⓐ	Ⓑ	Ⓒ	Ⓓ	Ⓔ	29.	Ⓐ	Ⓑ	Ⓒ	Ⓓ	Ⓔ
5.	Ⓐ	Ⓑ	Ⓒ	Ⓓ	Ⓔ	30.	Ⓐ	Ⓑ	Ⓒ	Ⓓ	Ⓔ
6.	Ⓐ	Ⓑ	Ⓒ	Ⓓ	Ⓔ	31.	Ⓐ	Ⓑ	Ⓒ	Ⓓ	Ⓔ
7.	Ⓐ	Ⓑ	Ⓒ	Ⓓ	Ⓔ	32.	Ⓐ	Ⓑ	Ⓒ	Ⓓ	Ⓔ
8.	Ⓐ	Ⓑ	Ⓒ	Ⓓ	Ⓔ	33.	Ⓐ	Ⓑ	Ⓒ	Ⓓ	Ⓔ
9.	Ⓐ	Ⓑ	Ⓒ	Ⓓ	Ⓔ	34.	Ⓐ	Ⓑ	Ⓒ	Ⓓ	Ⓔ
10.	Ⓐ	Ⓑ	Ⓒ	Ⓓ	Ⓔ	35.	Ⓐ	Ⓑ	Ⓒ	Ⓓ	Ⓔ
11.	Ⓐ	Ⓑ	Ⓒ	Ⓓ	Ⓔ	36.	Ⓐ	Ⓑ	Ⓒ	Ⓓ	Ⓔ
12.	Ⓐ	Ⓑ	Ⓒ	Ⓓ	Ⓔ	37.	Ⓐ	Ⓑ	Ⓒ	Ⓓ	Ⓔ
13.	Ⓐ	Ⓑ	Ⓒ	Ⓓ	Ⓔ	38.	Ⓐ	Ⓑ	Ⓒ	Ⓓ	Ⓔ
14.	Ⓐ	Ⓑ	Ⓒ	Ⓓ	Ⓔ	39.	Ⓐ	Ⓑ	Ⓒ	Ⓓ	Ⓔ
15.	Ⓐ	Ⓑ	Ⓒ	Ⓓ	Ⓔ	40.	Ⓐ	Ⓑ	Ⓒ	Ⓓ	Ⓔ
16.	Ⓐ	Ⓑ	Ⓒ	Ⓓ	Ⓔ	41.	Ⓐ	Ⓑ	Ⓒ	Ⓓ	Ⓔ
17.	Ⓐ	Ⓑ	Ⓒ	Ⓓ	Ⓔ	42.	Ⓐ	Ⓑ	Ⓒ	Ⓓ	Ⓔ
18.	Ⓐ	Ⓑ	Ⓒ	Ⓓ	Ⓔ	43.	Ⓐ	Ⓑ	Ⓒ	Ⓓ	Ⓔ
19.	Ⓐ	Ⓑ	Ⓒ	Ⓓ	Ⓔ	44.	Ⓐ	Ⓑ	Ⓒ	Ⓓ	Ⓔ
20.	Ⓐ	Ⓑ	Ⓒ	Ⓓ	Ⓔ	45.	Ⓐ	Ⓑ	Ⓒ	Ⓓ	Ⓔ
21.	Ⓐ	Ⓑ	Ⓒ	Ⓓ	Ⓔ	46.	Ⓐ	Ⓑ	Ⓒ	Ⓓ	Ⓔ
22.	Ⓐ	Ⓑ	Ⓒ	Ⓓ	Ⓔ	47.	Ⓐ	Ⓑ	Ⓒ	Ⓓ	Ⓔ
23.	Ⓐ	Ⓑ	Ⓒ	Ⓓ	Ⓔ	48.	Ⓐ	Ⓑ	Ⓒ	Ⓓ	Ⓔ
24.	Ⓐ	Ⓑ	Ⓒ	Ⓓ	Ⓔ	49.	Ⓐ	Ⓑ	Ⓒ	Ⓓ	Ⓔ
25.	Ⓐ	Ⓑ	Ⓒ	Ⓓ	Ⓔ	50.	Ⓐ	Ⓑ	Ⓒ	Ⓓ	Ⓔ

Model Test 10 Answer Sheet

#						#					
1.	Ⓐ	Ⓑ	Ⓒ	Ⓓ	Ⓔ	26.	Ⓐ	Ⓑ	Ⓒ	Ⓓ	Ⓔ
2.	Ⓐ	Ⓑ	Ⓒ	Ⓓ	Ⓔ	27.	Ⓐ	Ⓑ	Ⓒ	Ⓓ	Ⓔ
3.	Ⓐ	Ⓑ	Ⓒ	Ⓓ	Ⓔ	28.	Ⓐ	Ⓑ	Ⓒ	Ⓓ	Ⓔ
4.	Ⓐ	Ⓑ	Ⓒ	Ⓓ	Ⓔ	29.	Ⓐ	Ⓑ	Ⓒ	Ⓓ	Ⓔ
5.	Ⓐ	Ⓑ	Ⓒ	Ⓓ	Ⓔ	30.	Ⓐ	Ⓑ	Ⓒ	Ⓓ	Ⓔ
6.	Ⓐ	Ⓑ	Ⓒ	Ⓓ	Ⓔ	31.	Ⓐ	Ⓑ	Ⓒ	Ⓓ	Ⓔ
7.	Ⓐ	Ⓑ	Ⓒ	Ⓓ	Ⓔ	32.	Ⓐ	Ⓑ	Ⓒ	Ⓓ	Ⓔ
8.	Ⓐ	Ⓑ	Ⓒ	Ⓓ	Ⓔ	33.	Ⓐ	Ⓑ	Ⓒ	Ⓓ	Ⓔ
9.	Ⓐ	Ⓑ	Ⓒ	Ⓓ	Ⓔ	34.	Ⓐ	Ⓑ	Ⓒ	Ⓓ	Ⓔ
10.	Ⓐ	Ⓑ	Ⓒ	Ⓓ	Ⓔ	35.	Ⓐ	Ⓑ	Ⓒ	Ⓓ	Ⓔ
11.	Ⓐ	Ⓑ	Ⓒ	Ⓓ	Ⓔ	36.	Ⓐ	Ⓑ	Ⓒ	Ⓓ	Ⓔ
12.	Ⓐ	Ⓑ	Ⓒ	Ⓓ	Ⓔ	37.	Ⓐ	Ⓑ	Ⓒ	Ⓓ	Ⓔ
13.	Ⓐ	Ⓑ	Ⓒ	Ⓓ	Ⓔ	38.	Ⓐ	Ⓑ	Ⓒ	Ⓓ	Ⓔ
14.	Ⓐ	Ⓑ	Ⓒ	Ⓓ	Ⓔ	39.	Ⓐ	Ⓑ	Ⓒ	Ⓓ	Ⓔ
15.	Ⓐ	Ⓑ	Ⓒ	Ⓓ	Ⓔ	40.	Ⓐ	Ⓑ	Ⓒ	Ⓓ	Ⓔ
16.	Ⓐ	Ⓑ	Ⓒ	Ⓓ	Ⓔ	41.	Ⓐ	Ⓑ	Ⓒ	Ⓓ	Ⓔ
17.	Ⓐ	Ⓑ	Ⓒ	Ⓓ	Ⓔ	42.	Ⓐ	Ⓑ	Ⓒ	Ⓓ	Ⓔ
18.	Ⓐ	Ⓑ	Ⓒ	Ⓓ	Ⓔ	43.	Ⓐ	Ⓑ	Ⓒ	Ⓓ	Ⓔ
19.	Ⓐ	Ⓑ	Ⓒ	Ⓓ	Ⓔ	44.	Ⓐ	Ⓑ	Ⓒ	Ⓓ	Ⓔ
20.	Ⓐ	Ⓑ	Ⓒ	Ⓓ	Ⓔ	45.	Ⓐ	Ⓑ	Ⓒ	Ⓓ	Ⓔ
21.	Ⓐ	Ⓑ	Ⓒ	Ⓓ	Ⓔ	46.	Ⓐ	Ⓑ	Ⓒ	Ⓓ	Ⓔ
22.	Ⓐ	Ⓑ	Ⓒ	Ⓓ	Ⓔ	47.	Ⓐ	Ⓑ	Ⓒ	Ⓓ	Ⓔ
23.	Ⓐ	Ⓑ	Ⓒ	Ⓓ	Ⓔ	48.	Ⓐ	Ⓑ	Ⓒ	Ⓓ	Ⓔ
24.	Ⓐ	Ⓑ	Ⓒ	Ⓓ	Ⓔ	49.	Ⓐ	Ⓑ	Ⓒ	Ⓓ	Ⓔ
25.	Ⓐ	Ⓑ	Ⓒ	Ⓓ	Ⓔ	50.	Ⓐ	Ⓑ	Ⓒ	Ⓓ	Ⓔ

Model Test 11 Answer Sheet

No.	A	B	C	D	E		No.	A	B	C	D	E
1.	Ⓐ	Ⓑ	Ⓒ	Ⓓ	Ⓔ		26.	Ⓐ	Ⓑ	Ⓒ	Ⓓ	Ⓔ
2.	Ⓐ	Ⓑ	Ⓒ	Ⓓ	Ⓔ		27.	Ⓐ	Ⓑ	Ⓒ	Ⓓ	Ⓔ
3.	Ⓐ	Ⓑ	Ⓒ	Ⓓ	Ⓔ		28.	Ⓐ	Ⓑ	Ⓒ	Ⓓ	Ⓔ
4.	Ⓐ	Ⓑ	Ⓒ	Ⓓ	Ⓔ		29.	Ⓐ	Ⓑ	Ⓒ	Ⓓ	Ⓔ
5.	Ⓐ	Ⓑ	Ⓒ	Ⓓ	Ⓔ		30.	Ⓐ	Ⓑ	Ⓒ	Ⓓ	Ⓔ
6.	Ⓐ	Ⓑ	Ⓒ	Ⓓ	Ⓔ		31.	Ⓐ	Ⓑ	Ⓒ	Ⓓ	Ⓔ
7.	Ⓐ	Ⓑ	Ⓒ	Ⓓ	Ⓔ		32.	Ⓐ	Ⓑ	Ⓒ	Ⓓ	Ⓔ
8.	Ⓐ	Ⓑ	Ⓒ	Ⓓ	Ⓔ		33.	Ⓐ	Ⓑ	Ⓒ	Ⓓ	Ⓔ
9.	Ⓐ	Ⓑ	Ⓒ	Ⓓ	Ⓔ		34.	Ⓐ	Ⓑ	Ⓒ	Ⓓ	Ⓔ
10.	Ⓐ	Ⓑ	Ⓒ	Ⓓ	Ⓔ		35.	Ⓐ	Ⓑ	Ⓒ	Ⓓ	Ⓔ
11.	Ⓐ	Ⓑ	Ⓒ	Ⓓ	Ⓔ		36.	Ⓐ	Ⓑ	Ⓒ	Ⓓ	Ⓔ
12.	Ⓐ	Ⓑ	Ⓒ	Ⓓ	Ⓔ		37.	Ⓐ	Ⓑ	Ⓒ	Ⓓ	Ⓔ
13.	Ⓐ	Ⓑ	Ⓒ	Ⓓ	Ⓔ		38.	Ⓐ	Ⓑ	Ⓒ	Ⓓ	Ⓔ
14.	Ⓐ	Ⓑ	Ⓒ	Ⓓ	Ⓔ		39.	Ⓐ	Ⓑ	Ⓒ	Ⓓ	Ⓔ
15.	Ⓐ	Ⓑ	Ⓒ	Ⓓ	Ⓔ		40.	Ⓐ	Ⓑ	Ⓒ	Ⓓ	Ⓔ
16.	Ⓐ	Ⓑ	Ⓒ	Ⓓ	Ⓔ		41.	Ⓐ	Ⓑ	Ⓒ	Ⓓ	Ⓔ
17.	Ⓐ	Ⓑ	Ⓒ	Ⓓ	Ⓔ		42.	Ⓐ	Ⓑ	Ⓒ	Ⓓ	Ⓔ
18.	Ⓐ	Ⓑ	Ⓒ	Ⓓ	Ⓔ		43.	Ⓐ	Ⓑ	Ⓒ	Ⓓ	Ⓔ
19.	Ⓐ	Ⓑ	Ⓒ	Ⓓ	Ⓔ		44.	Ⓐ	Ⓑ	Ⓒ	Ⓓ	Ⓔ
20.	Ⓐ	Ⓑ	Ⓒ	Ⓓ	Ⓔ		45.	Ⓐ	Ⓑ	Ⓒ	Ⓓ	Ⓔ
21.	Ⓐ	Ⓑ	Ⓒ	Ⓓ	Ⓔ		46.	Ⓐ	Ⓑ	Ⓒ	Ⓓ	Ⓔ
22.	Ⓐ	Ⓑ	Ⓒ	Ⓓ	Ⓔ		47.	Ⓐ	Ⓑ	Ⓒ	Ⓓ	Ⓔ
23.	Ⓐ	Ⓑ	Ⓒ	Ⓓ	Ⓔ		48.	Ⓐ	Ⓑ	Ⓒ	Ⓓ	Ⓔ
24.	Ⓐ	Ⓑ	Ⓒ	Ⓓ	Ⓔ		49.	Ⓐ	Ⓑ	Ⓒ	Ⓓ	Ⓔ
25.	Ⓐ	Ⓑ	Ⓒ	Ⓓ	Ⓔ		50.	Ⓐ	Ⓑ	Ⓒ	Ⓓ	Ⓔ

Model Test 12 Answer Sheet

No.	A	B	C	D	E		No.	A	B	C	D	E
1.	Ⓐ	Ⓑ	Ⓒ	Ⓓ	Ⓔ		26.	Ⓐ	Ⓑ	Ⓒ	Ⓓ	Ⓔ
2.	Ⓐ	Ⓑ	Ⓒ	Ⓓ	Ⓔ		27.	Ⓐ	Ⓑ	Ⓒ	Ⓓ	Ⓔ
3.	Ⓐ	Ⓑ	Ⓒ	Ⓓ	Ⓔ		28.	Ⓐ	Ⓑ	Ⓒ	Ⓓ	Ⓔ
4.	Ⓐ	Ⓑ	Ⓒ	Ⓓ	Ⓔ		29.	Ⓐ	Ⓑ	Ⓒ	Ⓓ	Ⓔ
5.	Ⓐ	Ⓑ	Ⓒ	Ⓓ	Ⓔ		30.	Ⓐ	Ⓑ	Ⓒ	Ⓓ	Ⓔ
6.	Ⓐ	Ⓑ	Ⓒ	Ⓓ	Ⓔ		31.	Ⓐ	Ⓑ	Ⓒ	Ⓓ	Ⓔ
7.	Ⓐ	Ⓑ	Ⓒ	Ⓓ	Ⓔ		32.	Ⓐ	Ⓑ	Ⓒ	Ⓓ	Ⓔ
8.	Ⓐ	Ⓑ	Ⓒ	Ⓓ	Ⓔ		33.	Ⓐ	Ⓑ	Ⓒ	Ⓓ	Ⓔ
9.	Ⓐ	Ⓑ	Ⓒ	Ⓓ	Ⓔ		34.	Ⓐ	Ⓑ	Ⓒ	Ⓓ	Ⓔ
10.	Ⓐ	Ⓑ	Ⓒ	Ⓓ	Ⓔ		35.	Ⓐ	Ⓑ	Ⓒ	Ⓓ	Ⓔ
11.	Ⓐ	Ⓑ	Ⓒ	Ⓓ	Ⓔ		36.	Ⓐ	Ⓑ	Ⓒ	Ⓓ	Ⓔ
12.	Ⓐ	Ⓑ	Ⓒ	Ⓓ	Ⓔ		37.	Ⓐ	Ⓑ	Ⓒ	Ⓓ	Ⓔ
13.	Ⓐ	Ⓑ	Ⓒ	Ⓓ	Ⓔ		38.	Ⓐ	Ⓑ	Ⓒ	Ⓓ	Ⓔ
14.	Ⓐ	Ⓑ	Ⓒ	Ⓓ	Ⓔ		39.	Ⓐ	Ⓑ	Ⓒ	Ⓓ	Ⓔ
15.	Ⓐ	Ⓑ	Ⓒ	Ⓓ	Ⓔ		40.	Ⓐ	Ⓑ	Ⓒ	Ⓓ	Ⓔ
16.	Ⓐ	Ⓑ	Ⓒ	Ⓓ	Ⓔ		41.	Ⓐ	Ⓑ	Ⓒ	Ⓓ	Ⓔ
17.	Ⓐ	Ⓑ	Ⓒ	Ⓓ	Ⓔ		42.	Ⓐ	Ⓑ	Ⓒ	Ⓓ	Ⓔ
18.	Ⓐ	Ⓑ	Ⓒ	Ⓓ	Ⓔ		43.	Ⓐ	Ⓑ	Ⓒ	Ⓓ	Ⓔ
19.	Ⓐ	Ⓑ	Ⓒ	Ⓓ	Ⓔ		44.	Ⓐ	Ⓑ	Ⓒ	Ⓓ	Ⓔ
20.	Ⓐ	Ⓑ	Ⓒ	Ⓓ	Ⓔ		45.	Ⓐ	Ⓑ	Ⓒ	Ⓓ	Ⓔ
21.	Ⓐ	Ⓑ	Ⓒ	Ⓓ	Ⓔ		46.	Ⓐ	Ⓑ	Ⓒ	Ⓓ	Ⓔ
22.	Ⓐ	Ⓑ	Ⓒ	Ⓓ	Ⓔ		47.	Ⓐ	Ⓑ	Ⓒ	Ⓓ	Ⓔ
23.	Ⓐ	Ⓑ	Ⓒ	Ⓓ	Ⓔ		48.	Ⓐ	Ⓑ	Ⓒ	Ⓓ	Ⓔ
24.	Ⓐ	Ⓑ	Ⓒ	Ⓓ	Ⓔ		49.	Ⓐ	Ⓑ	Ⓒ	Ⓓ	Ⓔ
25.	Ⓐ	Ⓑ	Ⓒ	Ⓓ	Ⓔ		50.	Ⓐ	Ⓑ	Ⓒ	Ⓓ	Ⓔ

Model Test 13 Answer Sheet

1.	Ⓐ Ⓑ Ⓒ Ⓓ Ⓔ		26.	Ⓐ Ⓑ Ⓒ Ⓓ Ⓔ
2.	Ⓐ Ⓑ Ⓒ Ⓓ Ⓔ		27.	Ⓐ Ⓑ Ⓒ Ⓓ Ⓔ
3.	Ⓐ Ⓑ Ⓒ Ⓓ Ⓔ		28.	Ⓐ Ⓑ Ⓒ Ⓓ Ⓔ
4.	Ⓐ Ⓑ Ⓒ Ⓓ Ⓔ		29.	Ⓐ Ⓑ Ⓒ Ⓓ Ⓔ
5.	Ⓐ Ⓑ Ⓒ Ⓓ Ⓔ		30.	Ⓐ Ⓑ Ⓒ Ⓓ Ⓔ
6.	Ⓐ Ⓑ Ⓒ Ⓓ Ⓔ		31.	Ⓐ Ⓑ Ⓒ Ⓓ Ⓔ
7.	Ⓐ Ⓑ Ⓒ Ⓓ Ⓔ		32.	Ⓐ Ⓑ Ⓒ Ⓓ Ⓔ
8.	Ⓐ Ⓑ Ⓒ Ⓓ Ⓔ		33.	Ⓐ Ⓑ Ⓒ Ⓓ Ⓔ
9.	Ⓐ Ⓑ Ⓒ Ⓓ Ⓔ		34.	Ⓐ Ⓑ Ⓒ Ⓓ Ⓔ
10.	Ⓐ Ⓑ Ⓒ Ⓓ Ⓔ		35.	Ⓐ Ⓑ Ⓒ Ⓓ Ⓔ
11.	Ⓐ Ⓑ Ⓒ Ⓓ Ⓔ		36.	Ⓐ Ⓑ Ⓒ Ⓓ Ⓔ
12.	Ⓐ Ⓑ Ⓒ Ⓓ Ⓔ		37.	Ⓐ Ⓑ Ⓒ Ⓓ Ⓔ
13.	Ⓐ Ⓑ Ⓒ Ⓓ Ⓔ		38.	Ⓐ Ⓑ Ⓒ Ⓓ Ⓔ
14.	Ⓐ Ⓑ Ⓒ Ⓓ Ⓔ		39.	Ⓐ Ⓑ Ⓒ Ⓓ Ⓔ
15.	Ⓐ Ⓑ Ⓒ Ⓓ Ⓔ		40.	Ⓐ Ⓑ Ⓒ Ⓓ Ⓔ
16.	Ⓐ Ⓑ Ⓒ Ⓓ Ⓔ		41.	Ⓐ Ⓑ Ⓒ Ⓓ Ⓔ
17.	Ⓐ Ⓑ Ⓒ Ⓓ Ⓔ		42.	Ⓐ Ⓑ Ⓒ Ⓓ Ⓔ
18.	Ⓐ Ⓑ Ⓒ Ⓓ Ⓔ		43.	Ⓐ Ⓑ Ⓒ Ⓓ Ⓔ
19.	Ⓐ Ⓑ Ⓒ Ⓓ Ⓔ		44.	Ⓐ Ⓑ Ⓒ Ⓓ Ⓔ
20.	Ⓐ Ⓑ Ⓒ Ⓓ Ⓔ		45.	Ⓐ Ⓑ Ⓒ Ⓓ Ⓔ
21.	Ⓐ Ⓑ Ⓒ Ⓓ Ⓔ		46.	Ⓐ Ⓑ Ⓒ Ⓓ Ⓔ
22.	Ⓐ Ⓑ Ⓒ Ⓓ Ⓔ		47.	Ⓐ Ⓑ Ⓒ Ⓓ Ⓔ
23.	Ⓐ Ⓑ Ⓒ Ⓓ Ⓔ		48.	Ⓐ Ⓑ Ⓒ Ⓓ Ⓔ
24.	Ⓐ Ⓑ Ⓒ Ⓓ Ⓔ		49.	Ⓐ Ⓑ Ⓒ Ⓓ Ⓔ
25.	Ⓐ Ⓑ Ⓒ Ⓓ Ⓔ		50.	Ⓐ Ⓑ Ⓒ Ⓓ Ⓔ

Model Test 14 Answer Sheet

1.	Ⓐ Ⓑ Ⓒ Ⓓ Ⓔ		26.	Ⓐ Ⓑ Ⓒ Ⓓ Ⓔ
2.	Ⓐ Ⓑ Ⓒ Ⓓ Ⓔ		27.	Ⓐ Ⓑ Ⓒ Ⓓ Ⓔ
3.	Ⓐ Ⓑ Ⓒ Ⓓ Ⓔ		28.	Ⓐ Ⓑ Ⓒ Ⓓ Ⓔ
4.	Ⓐ Ⓑ Ⓒ Ⓓ Ⓔ		29.	Ⓐ Ⓑ Ⓒ Ⓓ Ⓔ
5.	Ⓐ Ⓑ Ⓒ Ⓓ Ⓔ		30.	Ⓐ Ⓑ Ⓒ Ⓓ Ⓔ
6.	Ⓐ Ⓑ Ⓒ Ⓓ Ⓔ		31.	Ⓐ Ⓑ Ⓒ Ⓓ Ⓔ
7.	Ⓐ Ⓑ Ⓒ Ⓓ Ⓔ		32.	Ⓐ Ⓑ Ⓒ Ⓓ Ⓔ
8.	Ⓐ Ⓑ Ⓒ Ⓓ Ⓔ		33.	Ⓐ Ⓑ Ⓒ Ⓓ Ⓔ
9.	Ⓐ Ⓑ Ⓒ Ⓓ Ⓔ		34.	Ⓐ Ⓑ Ⓒ Ⓓ Ⓔ
10.	Ⓐ Ⓑ Ⓒ Ⓓ Ⓔ		35.	Ⓐ Ⓑ Ⓒ Ⓓ Ⓔ
11.	Ⓐ Ⓑ Ⓒ Ⓓ Ⓔ		36.	Ⓐ Ⓑ Ⓒ Ⓓ Ⓔ
12.	Ⓐ Ⓑ Ⓒ Ⓓ Ⓔ		37.	Ⓐ Ⓑ Ⓒ Ⓓ Ⓔ
13.	Ⓐ Ⓑ Ⓒ Ⓓ Ⓔ		38.	Ⓐ Ⓑ Ⓒ Ⓓ Ⓔ
14.	Ⓐ Ⓑ Ⓒ Ⓓ Ⓔ		39.	Ⓐ Ⓑ Ⓒ Ⓓ Ⓔ
15.	Ⓐ Ⓑ Ⓒ Ⓓ Ⓔ		40.	Ⓐ Ⓑ Ⓒ Ⓓ Ⓔ
16.	Ⓐ Ⓑ Ⓒ Ⓓ Ⓔ		41.	Ⓐ Ⓑ Ⓒ Ⓓ Ⓔ
17.	Ⓐ Ⓑ Ⓒ Ⓓ Ⓔ		42.	Ⓐ Ⓑ Ⓒ Ⓓ Ⓔ
18.	Ⓐ Ⓑ Ⓒ Ⓓ Ⓔ		43.	Ⓐ Ⓑ Ⓒ Ⓓ Ⓔ
19.	Ⓐ Ⓑ Ⓒ Ⓓ Ⓔ		44.	Ⓐ Ⓑ Ⓒ Ⓓ Ⓔ
20.	Ⓐ Ⓑ Ⓒ Ⓓ Ⓔ		45.	Ⓐ Ⓑ Ⓒ Ⓓ Ⓔ
21.	Ⓐ Ⓑ Ⓒ Ⓓ Ⓔ		46.	Ⓐ Ⓑ Ⓒ Ⓓ Ⓔ
22.	Ⓐ Ⓑ Ⓒ Ⓓ Ⓔ		47.	Ⓐ Ⓑ Ⓒ Ⓓ Ⓔ
23.	Ⓐ Ⓑ Ⓒ Ⓓ Ⓔ		48.	Ⓐ Ⓑ Ⓒ Ⓓ Ⓔ
24.	Ⓐ Ⓑ Ⓒ Ⓓ Ⓔ		49.	Ⓐ Ⓑ Ⓒ Ⓓ Ⓔ
25.	Ⓐ Ⓑ Ⓒ Ⓓ Ⓔ		50.	Ⓐ Ⓑ Ⓒ Ⓓ Ⓔ

Model Test 15 Answer Sheet

1.	Ⓐ	Ⓑ	Ⓒ	Ⓓ	Ⓔ	
2.	Ⓐ	Ⓑ	Ⓒ	Ⓓ	Ⓔ	
3.	Ⓐ	Ⓑ	Ⓒ	Ⓓ	Ⓔ	
4.	Ⓐ	Ⓑ	Ⓒ	Ⓓ	Ⓔ	
5.	Ⓐ	Ⓑ	Ⓒ	Ⓓ	Ⓔ	
6.	Ⓐ	Ⓑ	Ⓒ	Ⓓ	Ⓔ	
7.	Ⓐ	Ⓑ	Ⓒ	Ⓓ	Ⓔ	
8.	Ⓐ	Ⓑ	Ⓒ	Ⓓ	Ⓔ	
9.	Ⓐ	Ⓑ	Ⓒ	Ⓓ	Ⓔ	
10.	Ⓐ	Ⓑ	Ⓒ	Ⓓ	Ⓔ	
11.	Ⓐ	Ⓑ	Ⓒ	Ⓓ	Ⓔ	
12.	Ⓐ	Ⓑ	Ⓒ	Ⓓ	Ⓔ	
13.	Ⓐ	Ⓑ	Ⓒ	Ⓓ	Ⓔ	
14.	Ⓐ	Ⓑ	Ⓒ	Ⓓ	Ⓔ	
15.	Ⓐ	Ⓑ	Ⓒ	Ⓓ	Ⓔ	
16.	Ⓐ	Ⓑ	Ⓒ	Ⓓ	Ⓔ	
17.	Ⓐ	Ⓑ	Ⓒ	Ⓓ	Ⓔ	
18.	Ⓐ	Ⓑ	Ⓒ	Ⓓ	Ⓔ	
19.	Ⓐ	Ⓑ	Ⓒ	Ⓓ	Ⓔ	
20.	Ⓐ	Ⓑ	Ⓒ	Ⓓ	Ⓔ	
21.	Ⓐ	Ⓑ	Ⓒ	Ⓓ	Ⓔ	
22.	Ⓐ	Ⓑ	Ⓒ	Ⓓ	Ⓔ	
23.	Ⓐ	Ⓑ	Ⓒ	Ⓓ	Ⓔ	
24.	Ⓐ	Ⓑ	Ⓒ	Ⓓ	Ⓔ	
25.	Ⓐ	Ⓑ	Ⓒ	Ⓓ	Ⓔ	

26.	Ⓐ	Ⓑ	Ⓒ	Ⓓ	Ⓔ	
27.	Ⓐ	Ⓑ	Ⓒ	Ⓓ	Ⓔ	
28.	Ⓐ	Ⓑ	Ⓒ	Ⓓ	Ⓔ	
29.	Ⓐ	Ⓑ	Ⓒ	Ⓓ	Ⓔ	
30.	Ⓐ	Ⓑ	Ⓒ	Ⓓ	Ⓔ	
31.	Ⓐ	Ⓑ	Ⓒ	Ⓓ	Ⓔ	
32.	Ⓐ	Ⓑ	Ⓒ	Ⓓ	Ⓔ	
33.	Ⓐ	Ⓑ	Ⓒ	Ⓓ	Ⓔ	
34.	Ⓐ	Ⓑ	Ⓒ	Ⓓ	Ⓔ	
35.	Ⓐ	Ⓑ	Ⓒ	Ⓓ	Ⓔ	
36.	Ⓐ	Ⓑ	Ⓒ	Ⓓ	Ⓔ	
37.	Ⓐ	Ⓑ	Ⓒ	Ⓓ	Ⓔ	
38.	Ⓐ	Ⓑ	Ⓒ	Ⓓ	Ⓔ	
39.	Ⓐ	Ⓑ	Ⓒ	Ⓓ	Ⓔ	
40.	Ⓐ	Ⓑ	Ⓒ	Ⓓ	Ⓔ	
41.	Ⓐ	Ⓑ	Ⓒ	Ⓓ	Ⓔ	
42.	Ⓐ	Ⓑ	Ⓒ	Ⓓ	Ⓔ	
43.	Ⓐ	Ⓑ	Ⓒ	Ⓓ	Ⓔ	
44.	Ⓐ	Ⓑ	Ⓒ	Ⓓ	Ⓔ	
45.	Ⓐ	Ⓑ	Ⓒ	Ⓓ	Ⓔ	
46.	Ⓐ	Ⓑ	Ⓒ	Ⓓ	Ⓔ	
47.	Ⓐ	Ⓑ	Ⓒ	Ⓓ	Ⓔ	
48.	Ⓐ	Ⓑ	Ⓒ	Ⓓ	Ⓔ	
49.	Ⓐ	Ⓑ	Ⓒ	Ⓓ	Ⓔ	
50.	Ⓐ	Ⓑ	Ⓒ	Ⓓ	Ⓔ	

"Shorten 40 hours of college preparatory precalculus study to an easy 4 hours…"

- The usage of the TI 83 – TI 84 family of graphing calculators particularly in the context of Algebra, Pre-Calculus and SAT and IB Mathematics with over 400 questions carefully designed and fully solved questions;
- A detailed additional chapter on TI 89 with a special focus on the algebra menu and the graphing facilities of this device;
- 10 original and fully solved sample tests 5 or Level 1 and 5 for Level 2.

This book is intended to help high school students who are bound to take either or both of the SAT Math Level 1 and Level 2 tests. Being one of a kind, this book is devoted to the usage of the TI 83 – TI 84 family of graphing calculators particularly in the context of Algebra, Pre-Calculus and SAT and IB Mathematics with over 400 questions carefully designed and fully solved questions. The method proposed in this book has been developed through 5 years' experience; has been proven to work and has created a success story each and every time, having helped hundreds of students who are currently attending the top 50 universities in the USA including many Ivy League schools.

The main advantage of the approach suggested in this book is that, one can solve, any equation or inequality with the TI, whether it is algebraic, trigonometric, exponential, logarithmic, polynomial or one that involves absolute values, without needing to know the related topic in depth and having to perform tedious steps One can solve all types of equations and inequalities very easily and in a very similar way just needing to learn a few very easy to remember techniques.

But there are still more to what can be done with the TI; find period, frequency, amplitude, offset, axis of wave of a periodic function, find the maxima minima and zeros as well as the domains and ranges of all types of functions; perform any operations on complex numbers, carry out any computation involving sequences and series, perform matrix algebra, solve a system of equations for any number of unknowns and even write small programs to ease your life. More than 20 of the 50 questions in the SAT Mathematics Subject Tests are based on the topics given above and this is why this book upgrades the SAT Mathematics subject test scores by at least 200 points.

Topics covered are:

- EQUATIONS: Polynomial, Algebraic, Absolute Value, Exponential and Logarithmic, Trigonometric, Inverse Trigonometric
- INEQUALITIES: Polynomial, Algebraic and Absolute Value, Trigonometric
- FUNCTIONS: Maxima and Minima, Domains and Ranges, Evenness and Oddness, Graphs of Trigonometric Functions, the Greatest Integer Function
- BASIC CALCULUS: Zeros, Holes, Limits, Continuity, Horizontal and Vertical Asymptotes
- CONIC SECTIONS: Circle, Ellipse, Parabola and Hyperbola
- LINEAR ALGEBRA: System of Linear Equations, Matrices and Determinants
- MISCELLANEOUS: Parametric and Polar Graphing; Complex Numbers; Permutations and Combinations, Computer Programs, Sequences and Series, Statistics, and more…

Complete Prep for the SAT Math Subject Tests

Level 1 and Level 2 with 10 Fully Solved Sample Tests

The only complete prep for the SAT Math Level 1 and Level 2 Subject Tests

The book covers every topic that has ever appeared on both tests in detail and accurately. Moreover there are 10 sample tests, 5 from each level.

The topics covered are: Basic algebra; number theory; equations; inequalities; unit conversions; logic; arithmetic, geometric and harmonic means; basic computer programs; operations; functions; evenness and oddness; basic functions; transformations; advanced graphing; symmetries; rotations; linear functions; quadratic functions; polynomial functions; trigonometry; exponentials and logarithms; limits, continuity and asymptotes; the greatest integer function; absolute values; conic sections; complex numbers; parametric and polar coordinates; variation; locus; plane and three dimensional geometry; inscribed figures (two or three dimensional); rotations in three dimensions; permutations, combinations and probability; binomials; sequences and series; statistics; matrices and determinants; three dimensional coordinate geometry; vectors; data analysis (graphs and tables) and all kinds of word problems.

About the Author

Ruşen* Meylani is the co – founder of **RUSH** Publications and Educational Consultancy, LLC. Born on the 2nd of August, 1972 he shows all the leadership characteristics of a Leo. He was awarded many times in mathematics since the age of 14 and he holds a Master of Science degree in Electrical and Electronics Engineering having written tens of papers in this field. When he was a graduate student (between 1995 and 1997), he invented three methods that are currently being used with sophisticated printing devices, the last generation white goods, and, the state of the art quality control systems. He worked in the field of Information Technology where he was the leader of a team that built the wide area network of a major supermarket chain in Europe.

In 1998 he decided to build a career in education listening to the sound of his heart and that was it. He created a method that shortens 40 hours of mathematics to 4 hours gaining the attraction of Eisenhower National Clearinghouse funded by the US Department of Education. In 2004 and 2005 he gave several conferences in the United States on this particular method. In the mean time he published three SAT II Mathematics books that became bestselling among their peers in Amazon just in a few months. He has 15 other books that will have all been published by mid 2007. He is a researcher, educator and academician having taught at several distinguished institutions at the K9 – 12, undergraduate and graduate levels, all in Europe. He has dedicated his life to creating easy to teach and easy to learn methods in mathematics. He considers himself as a "gifted loony" as he claims that one day it will be possible to shorten the learning time to less than one tenth of the usual. When people ask him about how it will be done, he borrows Albert Einstein's words: "If at first the idea is not absurd then there is no hope for it."

However, he is not an "all work, no play" type of person. He is a guitar player, poet and story writer being awarded several times in the United States for his poems. He is also a great movie watcher. He is happily married to the sweetest genius lady who is crazy enough to marry him. The couple has created their own language which is the funniest way of communication that has ever existed in the world.

Ruşen Meylani's motto is somewhat similar to Robert Kennedy's: "Some men see things as they are and say 'Why?' I see the things that never were and say 'Why not!' "

www.rusenmeylani.com / www.rushsociety.com

*ş is pronounced as "sh" like in "Ash Wednesday" (a poem by T.S. Elliot)

 **collegeboard.com**

SAT Registration & Scores

SAT Registration

Date	Test	Score	Percentile*	Status
10/2004	**SAT II**			Test Completed
	Math Level 1 w/Calculator	800	99	
	Math Level 2 w/Calculator	800	90	
10/2002	**SAT II**			Test Completed
	Math Level 1 w/Calculator	800	99	
	Physics	800	93	
05/2002	**SAT II**			Test Completed
	Math Level 1 w/Calculator	800	99	
	Math Level 2 w/Calculator	800	90	
	Physics	800	93	
06/2000	**SAT II**			Test Completed
	Math Level 1 w/Calculator	800	99	
	Math Level 2 w/Calculator	800	90	
	Physics	800	93	
05/2000	**SAT II**			Test Completed
	Math Level 1 w/Calculator	800	99	
	Math Level 2 w/Calculator	800	90	